# STATISTICAL METHODS FOR CANCER STUDIES

# STATISTICS: Textbooks and Monographs

## A SERIES EDITED BY

### D. B. OWEN, Coordinating Editor
*Department of Statistics*
*Southern Methodist University*
*Dallas, Texas*

### R. G. CORNELL, Associate Editor for Biostatistics
*School of Public Health*
*University of Michigan*
*Ann Arbor, Michigan*

## OTHER VOLUMES IN PREPARATION

# STATISTICAL METHODS FOR CANCER STUDIES

edited by
## RICHARD G. CORNELL

Department of Biostatistics
School of Public Health
The University of Michigan
Ann Arbor, Michigan

CRC Press
Taylor & Francis Group
Boca Raton London New York

CRC Press is an imprint of the
Taylor & Francis Group, an **informa** business

First published 1984 by Marcel Dekker, Inc.

Published 2019 by CRC Press
Taylor & Francis Group
6000 Broken Sound Parkway NW, Suite 300
Boca Raton, FL 33487-2742

© 1984 by Taylor & Francis Group, LLC
CRC Press is an imprint of Taylor & Francis Group, an Informa business

First issued in paperback 2019

No claim to original U.S. Government works

ISBN 13: 978-0-367-45186-8 (pbk)
ISBN 13: 978-0-8247-7169-0 (hbk)

**Visit the Taylor & Francis Web site at
http://www.taylorandfrancis.com**

**and the CRC Press Web site at
http://www.crcpress.com**

Library of Congress Cataloging in Publication Data
Main entry under title:

Statistical methods for cancer studies.

   (Statistics, textbooks and monographs ; v. 51)
   Includes index.
   1.  Cancer--Research--Statistical methods.
2.  Epidemiology--Statistical methods.  I. Cornell,
Richard G.  II. Series.  [DNLM:  1. Statistics.
2.  Neoplasms--Occurrence.  QZ 206 S7966]
RC262.S815  1984      616.99'4'0072      83-26319

# PREFACE

The prevention, detection and treatment of cancer increasingly have become major focuses of epidemiologic and medical research during the last decade. Included have been studies of possible risk factors which predispose cancer, including genetic, demographic, medical, life-style and environmental variables. There has also been an emphasis on the evaluation of programs for screening for disease, of regimens for treatment, and of experiments on the carcinogenic potential of compounds.

The evaluation of treatment regimens is not a major focus of this book. The primary methods for this purpose, randomized clinical trials and survival analysis, are discussed extensively, but the former is presented in the

context of the evaluation of screening programs and the
latter is developed more fully in the context of studies of
animal carcinogenesis.  Thus the emphasis is on public
health and epidemiologic aspects of cancer instead of
medical aspects.  A discussion of clinical trials that does
focus on the evaluation of alternative treatments is given
by Shapiro and Louis (1983), *Clinical Trials: Issues and
Approaches*, Marcel Dekker, Inc., New York.

The specialized nature of cancer studies has led to the
concurrent development of statistical methodology.  The
purpose of this book is to describe this methodology.  It is
aimed at readers familiar with the common concepts and
methods of statistics, including basic probability, topics
in statistical inference, such as maximum likelihood
estimation, and basic approaches to data analysis, such as
regression analysis and the analysis of variance.  Such
topics are often referred to but not presented in detail.
However, more advanced or more specialized procedures are
described more completely.

The rapid development of statistical methods for cancer
studies as reflected by a large number of papers in the
statistical literature on the topics covered in this book
has created a need for this reference for individual study
and for specialized courses.  These topics will be further
developed in the near future, so this book presents a picture
of a dynamic process which has not been completed.  Yet this
picture clarifies important features which can be studied

further through the literature which is summarized and
referenced.

   I am very grateful to the authors who have contributed
to this volume for their patience as well as for their
thorough presentations. The articles were prepared over a
time span of more then two years which necessitated review
and revision during the last year, particularly for those
chapters which were completed most promptly initially. Many
individuals who have contributed to the volume are
acknowledged in the separate chapters. I would particularly
like to thank Elizabeth A. Griffin, who has entered the
entire manuscript into the word processing system at the
University of Michigan and diligently attended to
modifications as the need for them has arisen.

                                          Richard G. Cornell

# CONTENTS

# CONTRIBUTORS

NORMAN E. BRESLOW, Ph.D., Department of Biostatistics and
Fred Hutchinson Cancer Research Center, University
of Washington, Seattle, Washington

WILLIAM J. BUTLER, Ph.D., Department of Biostatistics,
University of Michigan, Ann Arbor, Michigan

RICHARD G. CORNELL, Ph.D., Department of Biostatistics,
University of Michigan, Ann Arbor, Michigan

NICHOLAS E. DAY, Ph.D., Unit of Biostatistics, International
Agency for Research on Cancer, Lyon, France

DIANNE M. FINKELSTEIN, Ph.D., Biostatistics Department,
Harvard University School of Public Health and
Sidney Farber Cancer Institute, Boston,
Massachusetts

JOHN J. GART, PH.D., Mathematical Statistics and Applied
Mathematics Section, National Cancer Institute,
Chevy Chase, Maryland

JUN-MO NAM, M.S., Mathematical Statistics and Applied
Mathematics Section, Biometry Branch, National
Cancer Institute, Bethesda, Maryland

ALONZO L. PLOUGH, Ph.D., Department of Urban and
        Environmental Policy, Tufts University, Medford,
        Massachusetts

PHILIP C. PROROK, Ph.D., Biometrics and Operations Research
        Branch, National Cancer Institute, Bethesda,
        Maryland

MICHAEL SHWARTZ, Ph.D., Health Care Management Program,
        School of Management, Boston University, Boston,
        Massachusetts

MICHAEL J. SYMONS, Ph.D., Department of Biostatistics and
        Occupational Health Studies Group, University of
        North Carolina at Chapel Hill, Chapel Hill, North
        Carolina

JOHN D. TAULBEE, Ph.D., Miami Valley Laboratories, The
        Procter and Gamble Company, Cincinnati, Ohio

DONOVAN J. THOMPSON, Ph.D., Department of Biostatistics,
        University of Washington, Seattle, Washington

GEORGE W. WILLIAMS, Ph.D., Department of Biostatistics, The
        Cleveland Clinic Foundation, Cleveland, Ohio

ROBERT A. WOLFE, Ph.D., Department of Biostatistics,
        University of Michigan, Ann Arbor, Michigan

# STATISTICAL METHODS
# FOR CANCER STUDIES

# 1
# ASSESSING THE OCCURRENCE
# OF CANCER IN HUMAN POPULATIONS

Donovan J. Thompson
Department of Biostatistics
University of Washington
Seattle, Washington

## I.  INTRODUCTION

The bulk of the material in this book describes statistical
methods useful in studying problems arising in the field of
cancer.  The methods assume that human cancer can be
sufficiently well defined (1) to identify a case of one of
these diseases when it occurs in a population; (2) to permit
assignment of mortality due to cancer to one of the cancer
rubrics with sufficient accuracy to be useful; or (3) to
provide unambiguous operational time of occurrence data.
Whether these assumptions are justified in a particular
application of the methods will depend upon a number of
circumstances.  One purpose of this chapter will be to

describe some of the characteristics of these diseases which
have particular relevance to these circumstances. A second
objective of the chapter will be to set forth in some detail
the sources of information concerning the frequency of
occurrence of human cancer. So-called tumor registries are
an important element in this history and a forerunner of
current large data gathering operations in this field.
Consequently registries and their sequallae will be
described in a separate section. A final section will
describe the strengths and shortcomings of available
information on cancer relative to the purpose of this book.

## II.   IMPORTANT CHARACTERISTICS OF THESE DISEASES

Malignant neoplasms (in lay terms potentially life
threatening abnormal growth) apparently can arise in any
body tissue. In classifying them it is customary to relate
them to the organ in which they arise (the so-called primary
site) or the body system affected as in the leukemias and
lymphomas. In addition to local invasion some have the
capacity to spread to other sites in the body via the lymph
or circulatory system, the ability to metastasize. This
characteristic has led to the development of various systems
attempting to classify the extent of the disease process at
a point in time, say at diagnosis. These classifications
range from simple categorizations such as localized,
regional and distant spread, to more complicated schemes.
New staging schemes are continually being developed and

tested with recent emphasis shifting to the development of schemes that are site specific.

Unfortunately, even for primary tumors of the same organ, there is wide variation in the rate at which the disease process proceeds. In studying this phenomenon pathologists have refined the classification of tumors of a particular site, e.g. breast, by utilizing the minute structure, composition and function of the affected tissue in their classification; the histological type of tumor. Tumors of different histological types have different prognoses and may have different etiologies. The histological classification has been very useful in helping to understand the variability in the rates of progression. Attempts have also been made to grade the degree of malignancy of a particular tumor from the appearance of the cells in the tumor, poorly differentiated to well differentiated, reflecting rapid uncontrolled growth to less rapid.

Recent developments in medical technology, as represented by markedly improved mammography, thermography, and CAT scanners have made possible the visualization of tumors that previously were undetectable, and have also aided clinicians in determining the extent of invasion of surrounding tissues. Coupled with the previously well known phenomenon of occult tumors (tumors that remain undiagnosed during the lifetime of the host; e.g. routine examination of the prostate in older men coming to autopsy will reveal

prostatic cancer which during life had not caused symptoms
or required medical intervention) these developments have
raised the probability of detecting tumors that would
otherwise never surface clinically during the lifetime of
the host.  Additionally, the lack of understanding of the
highly variable behavior of different kinds of tumors from
onset of the neoplastic process to clinical surfacing, which
has been termed the latent or induction period, raises all
sorts of questions and difficulties in investigating
potential carcinogenic (cancer producing) insults in humans.
For many tumors the latent period appears to be in excess of
20 years (vinyl chloride induced liver tumors as a good
example) while for others it may be quite short (endometrial
cancer and exogenous estrogen).  Moreover there is no reason
to believe that an exposure of a young person to a
carcinogen would have the same result as exposure of an old
person to the same insult.  And finally with respect to
carcinogenic exposure, the relationship between exposure and
tumor may or may not be dose and duration dependent.  There
is clear evidence of a dose response effect in tumor genesis
for cigarette smoking (U.S. Department of Health, Education,
and Welfare (1974)).  There is also evidence for the effect
of duration of constant exposure in the case of endometrial
cancer and exogenous estrogens (Weiss et al.(1979)).  For
other tumors the situation is not clear.  Our limited
understanding of the basic biology of the neoplastic
process, the events that initiate it, promote it or protect

against it, is giving rise to such terms as co-carcinogens, promoters, and anti-carcinogens, which simply recognize the complexity of the etiology of these diseases. All these facts and issues bear on the analysis problems utilizing the methodologies described later in this book.

An additional very interesting facet of these diseases is the extent of the variability in their occurrence by age, ethnicity, geography, urbanization, and similar factors. Tumors that are relatively common in anglo-American residents of New Mexico occur much less frequently in residents of Mexican or Hispanic origin (U.S. Department of Health and Human Services (1981)). Hong Kong Chinese have very different rates of nasopharyngeal cancer than British or southeast Asians residing in Hong Kong (Ho (1972)). Such variability may be in part genetic in origin as well as environmental, both personal and community. As will be described in more detail later, this variability complicates the comparison of cancer incidence in one population group with that in another.

Summarizing, the primary site, histological type, mode of detection and various characteristics of the host and his environments will require attention in statistical analyses of observational studies of cancer occurrence. Two reference volumes that can be recommended to the serious investigator with limited biological background are those edited by Fraumeni (1975) and by Schottenfeld and Fraumeni (1982).

III.  SOURCES OF INFORMATION ON HUMAN CANCER

In the United States, death certificates provided for many
years most of the information on the occurrence of cancer.
Later particular hospitals began compiling information on
cases treated in their facilities.  Some hospitals made
efforts to identify and follow-up all such cancer cases.
Beginning in the mid 1930's the state of Connecticut
instituted a statewide system of promoting the notion of
registration and follow-up as part of a program aimed at
improving the care of cancer patients.  A little later than
the Connecticut undertaking (1937) the federal government
undertook the first of three national cancer surveys aimed
at providing information on cancer incidence for
approximately 10% of the U.S. population.  These studies
were sufficiently detailed to permit calculation of
incidence rates by sex, race, and age of the patient and
anatomic site and histologic type of the cancer.  The third
study covering the three years bracketing the 1970 decennial
census (National Cancer Institute Monograph 41, U.S.
Department of Health, Education, and Welfare (1975))
remained the best available information on cancer incidence
for the U.S. until the beginning of the Surveillance,
Epidemiology, and End Results (SEER) Program of the National
Cancer Institute in 1973.  Building on the third national
survey and expanding to several new geographical areas, the
SEER operation is a population based cancer reporting system

that covers about 25 million residents of the United States
and Puerto Rico.  The information abstracted from the
hospital records of diagnosed cancer cases also provides
data on the extent of the disease at diagnosis and some
information on the first course of therapy.  Survival data
is also becoming available as time passes.  The areas
involved in the SEER program are Connecticut, Hawaii, Iowa,
New Mexico, the San Francisco Bay area, the Puget Sound area
of Washington, Metropolitan Detroit, several counties in
Georgia including Metropolitan Atlanta, and Puerto Rico.
The first publication of incidence data from this system is
the monumental National Cancer Institute Monograph 57
(U.S. Department of Health, Education, and Welfare (1981))
covering the period 1973 to 1977.

Survival information on cancer patients in the United
States has, until recent years, come mainly from the old End
Results Group of the Biometry Branch of the National Cancer
Institute.  Several monographs have been prepared, the most
recent of which (U.S. Department of Health, Education, and
Welfare (1976)) provides data covering the period 1950-1973.
Data from the End Results Group has the shortcoming of not
being population based.  Survival rates at 3 and 5 years
post diagnosis can now be calculated from the population
based SEER data, but thus far none of these rates have been
published.  Therapy information in this data set most likely
will be restricted to whether or not surgery and/or
radiation was utilized.

The most specific survival data in the sense of
controlling for therapy have been provided in recent years
through the publications emanating from the various
cooperative oncology groups in the United States.  Many of
these investigations are clinical trials of specific
therapies for particular tumors.  A good entree into the
work of these groups can be obtained by perusing the journal
Cancer Treatment Reports for recent years.  The progress in
fighting Wilms' tumor in children accomplished by the
Children's Cancer Study Group as detailed in the
publications of this group is one of remarkable improvement
in survival.  Somewhat similar progress has been
accomplished in other groups for particular tumors (non-
Hodgkins lymphoma and cancer of the testis), but for the
major cancer sites survival prospects have improved, at
best, very little in the past 25 years.  These facts bear on
the appropriateness of mortality or incidence as the
endpoint to be used in particular studies.

On the international level cancer registration
activities of various extents and quality are ongoing.  The
most recent Cancer Incidence In Five Continents (Waterhouse
et al. (1976)), published by the International Agency for
Research in Cancer is perhaps the best general reference.
Population based registries for entire countries have
existed for many years, e.g. Denmark, New Zealand.

IV.  TUMOR REGISTRIES

For a thoroughgoing account of cancer registration with a
distinctly international flavor, the interested reader is
referred to a recent, excellent publication in the series
from the International Agency For Research on Cancer
(Maclennan *et al.* (1978)).  For the purposes of this book it
will be sufficient to restrict descriptions to tumor
registries typical of U.S. experience.

The typical tumor registry, whether for a single
hospital or for population based operations such as SEER,
depends upon the information found in the patient's hospital
chart.  This record includes admitting information on the
patient, physician's notes and summaries, orders for
medication, description of diagnostic, pathologic and
operative procedures and findings, test results, nursing
notes and other information.  The record is built by
admitting clerk, physician, surgeon, consultants,
pathologist, pharmacist, lab technicians, residents, and
nurses, each following patterns of recording that have
developed over the years for various reasons--medical,
legal, professional, personal.  Following discharge of the
patient, the responsible physician prepares a discharge
summary and the record is delivered to the medical record
librarian for storage.

It is at this point in time that tumor registrars or
specially trained abstractors attempt to locate all the
information contained in the chart relevant to the items
shown on the abstract form.  One current SEER abstract form
is reproduced below.  From the brief description of the way
a hospital chart gets put together, it should be clear that
it is not an ideal data gathering instrument.  Additionally,
as the practice of medicine changes, with diagnostic
procedures formerly done in the hospital now being performed
in special clinics or physicians' offices, and with cancer
therapy (radiotherapy, chemotherapy, immunotherapy, and
hormonal therapy) being done on an ambulatory basis, the
hospital chart becomes only one of several records that
might be required to adequately describe the diagnostic
workup and the therapy received.  Moreover, referral
practices among physicians result in multiple hospitals
becoming involved in the diagnostic and therapeutic process
for a single patient.  The records of two or more
hospitalizations would need to be abstracted and collated to
provide the data for such a patient.

Nevertheless the abstract is prepared and enters the
coding and editing process.  The example in Figure 1 is an
abstract of a patient with lung cancer.

The handwritten information has been summarized from
perusal of the hospitalization record by a specially trained
medical record abstractor.  The patient's name, address, and
other identifying information have been obliterated to

| DATE: 5-23-78 | | | | | | Cen. Reg. No. | ⓪ 0 1 7 2 5 7 | Ck.D. | 0 |
| BY: Ewhite | | | | CANCER SURVEILLANCE SYSTEM | | | | | |
| Code Ewhite  Edit KE | | | CONFIDENTIAL REPORT OF NEOPLASM | | | | File No. | | 2 |

**PATIENT NAME:** LAST    FIRST    MIDDLE ⓪

| TR # 77-184 | AGE 53 | SEX M | MARITAL STATUS M | RACE/ETHNICITY CAUC. | | | | Sex ① I | Marital Status ② 2 | Race ⓪ 0 |
| ADDRESS | STREET | CITY | STATE | | BIRTH PLACE CT. | | Birthplace ⓪ 0 0 7 | County ① 7 |

| Was This Diagnosis Made Elsewhere? ☐ Yes: Date    ☒ No | ITEMS BELOW IN BOLD TYPE TO BE COMPLETED AT HOSPITAL |
| Hospital (if yes) | Hospital Number ⓪ |
| SEQUENCE  ONE | SS No. ① |
| PRIMARY SITE  LUNG, RUL | Zip Code ⑨ 9 8 1 3 3 |

**PATHOLOGY** 12-29-77· Dx: RUL, LUNG WITH TWO SEPARATE NEOPLASMS. 1) A WELL-DIFF. ADENOCA. (SCAR CARCINOMA) 2) UNDIFF. LARGE CELL CARCINOMA. 2/3 PARATRACHEAL LYMPH NODES WITH METASTATIC UNDIFF. CA. GROSS IN MIDDLE OF LUNG, AT THE PERIPHERY, ABOUT 1 CM FROM PLEURAL SURFACE, A TUMOR 1 X 1 CM. AT BASE OF LOBE, A SCARRED AREA, 0.7 X 0.5 X 0.5 CM. WITH DIMPLING OF PLEURAL SURFACE. FS TAKEN FROM PERIPHERALLY LOCATED TUMOR MASS.

| Birthdate ⑩ 1 0 1 3 2 4 |
| Chart No. ⑫ 3 6 6 8 1 9 |
| Class of Case I | Date Admit ① 1 2 7 7 |
| Sequence ⑬ 1 2 7 7 |
| Date Dx ⑭ 1 2 7 7 |
| Primary Site ⑮ 6 2 3 |
| Histology ⑯ 8 1 4 0 3 1 |
| Dx Conf. ⑰ 1 |
| Dx Proc. ⑱ 0 0 0 6 |

| DATE | PROCEDURE | DATE | PROCEDURE |
| | Physical Examination 12-14-77. REF. FROM SAND POINT WITH LESION ON ROUTINE CHEST X-RAY. SMOKED 2 PPD X 37 YRS. QUIT 2 YRS. AGO. MILD COPD. HAS GUIAC + STOOLS. NO SOB, DYSPNEA, HEMOPTYSIS, WT. LOSS, NIGHT SWEATS, HAS FATIGUE, HBP. HEENT NEG. ABD-SL.OBESE. NO ADENOPATHY-NECK, GROIN AXILLA. | | Endoscopy 12-27-77· BRONCHOSCOPY· NL. AIRWAYS, CORDS NL. BXS. TAKEN |
| 12-14-77 | X-ray, scans CHEST· 1½-2 CM DENSITY RUL, ↑ IN SIZE SINCE 11-76. | | Surgical Observation 12-28-77· THORACOTOMY WITH RT. UPPER LOBECTOMY· MASS IN MID PORTION OF LUNG, GROSSLY + NODES IN RT GREATER FISSURE AND INTER-HILAR AREA, ANT. PRETRACHEAL AREA. |
| 12-28-77 | BONE SCAN. NL. | | |
| 12-27-77 | BE · NL. | | |
| 12-28-77 | UGI · HIATAL HERNIA, OTHER-WISE NL. | | |
| 12-27-77 | Laboratory ALK PHOS· NL. | | |

| | | 1 | 4 | 7 | 10 | 13 |
| FIRST COURSE DEFINITIVE THERAPY | | EOD ① 0 7 - 2 1 0 0 0 0 0 0 |
| 12-27-77· BEGAN "LUNG" DOUBLE BLIND STUDY- BCG, LEVAMISOLE, INH | Stage ② 1 1 | Laterality ① 1 | Mult ① 5 |
| 12-28-77· RU LOBECTOMY TO HAVE RADIATION RX. | Date 1st Rx ② 1 2 7 7 |
| | First Course Rx ⓪ 1 1 3 0 0 0 0 |
| STATUS AT ☒ ALIVE ☐ DEAD  DATE: 01-06-78 | Date last followed ④ 0 1 7 8 |
| LAST CONTACT ☐ Free of Ca ☐ With Ca ☒ Unknown Status | Status ④ 3 | FU Interval ④ 1 |
| PHYSICIAN FOLLOWING PATIENT: Name: Address: | Physician No. ⑤ 0 7 8 8 5 |
| DATE KEYTAPED: | Date Abst. ⑥ 0 5 7 8 | Abstractor 2 9 |
| N-5 REV. 6/79 ⑧ | | 0 6 7 9 |

Figure 1    Typical tumor registry abstract for case of lung cancer.

preserve anonymity, but otherwise the abstract is real. By
examining the abstract it may be deduced that the patient
was seen by a physician in a small town in 1976 where a
chest x-ray and guiac stool test suggested the need for
further workup. The patient was transferred later to a
larger urban hospital where he was examined on December 14,
1977 and the suspicious lung mass confirmed by x-ray.
Additional tests were performed, followed by a thorocotomy
on December 28 with the accompanying pathology report dated
December 29. Apparently the patient was entered on a
protocol study in the urban hospital and arrangements were
made to have the radiation therapy specified in the protocol
given at another medical facility. The patient was
discharged alive from the urban hospital on January 6, 1978.

To assist in getting the information on the abstract
into electronic data processing equipment, codes have to be
devised for the non-quantitative information. The items
coded from the abstract are identified on the right margin
of the form. The codes, together with a set of rules
governing their application, are a key element in attempting
to achieve reproducibility and uniformity over both time and
place in the data reduction process. The most complete set
of these procedures for cancer has been prepared by the SEER
program office and by the Statistical and Quality Control
Center (SAQC) for the Centralized Cancer Patient Data System
(CCPDS) (Fred Hutchinson Cancer Research Center (1981)).
Both of these systems rely heavily on the International

Classification of Disease-Oncology (ICD-O) for
classification of tumor topography and histology and the
International Classification of Disease Adapted (ICDA) for
classification of causes of death. Both ICD-O and ICDA are
the most recent of a series of disease classification
manuals developed through international cooperation under
the International Institute of Statistics and WHO auspices.

When an abstract has been coded and entered into an
electronic data processing machine, it can be subjected to a
wide variety of automatic editing checks designed to flag
unlikely codes and inconsistent codes in two or more fields.
These software programs are available for the data set shown
in the abstract from the SEER program or SAQC.

Summarizing, tumor registries are in operation in many
hospitals worldwide. In the United States they form the
basis for a large population based incidence and survival
reporting system (SEER) and a cooperative national patient
data system for the Comprehensive Cancer Research Centers
(CCPDS-SAQC). As the years pass these systems may be
expected to produce incidence and survival information by
age, sex, ethnicity, and other characteristics of the
patient and primary site, histology of the tumor, and extent
of disease at diagnosis that will provide one standard
against which to judge the occurrence and survival from
these diseases in special groups of persons. Useful
information on response to therapy is likely to be crude
from these systems and this may remain the case
indefinitely.

V.   STRENGTHS AND WEAKNESSES OF AVAILABLE CANCER DATA

Large portions of the world are covered by mortality
reporting systems that provide cause of death information.
Cancer has long been of interest as a major cause of
mortality worldwide.  In the United States mortality
reporting is essentially complete and publications are
reasonably current.  The data is, of course, available for
local areas by characteristics of importance such as age and
sex.  These are strengths.  The SEER system provides
incidence reporting that in addition to being population
based and complete for the area covered, contains additional
important information not available in the mortality data.
Among these items are histological type of the tumor, extent
of disease at diagnosis, and survival time postdiagnosis.
The CCPDS-SAQC system provides similar information to that
in SEER for an important set of treatment facilities in the
United States. Similar facilities in some other parts of the
world provide population based incidence and survival data
through their tumor registries since all care provided for
cases of cancer diagnosed in these areas is given through
these institutions.  This is not the case in the United
States as well as in other large parts of the world.
Nevertheless, it is an acknowledged fact that for no other
major disease entity is the currently available incidence
and mortality information as great as that for cancer.

       Given that extensive data on cancer incidence and
mortality is available, what do we know about its quality

and accuracy? In an earlier section of this chapter we have made some comments concerning the limitations of the basic sources of most data on cancer occurrence and survival. There have been several studies of the agreement between cause of death in U.S. vital records and information in the medical record of the deceased (Gittlesohn and Senning (1981), Percy *et al.* (1981)). Gittlesohn and Senning found in Vermont that for 17% of the death certificates with a specific cancer as the cause of death the hospital record showed that the cancer was of a different site or there was no mention of a neoplasm in the hospital record.

No extensive carefully designed study of the data produced by tumor registries has as yet been conducted. However, an example of a well planned and conducted small study is one undertaken by 18 comprehensive cancer centers in the United States under the auspices of SAQC (Feigl *et al.* (1983). Each of the 18 centers coded a common set of 25 abstracts selected by SAQC as typical for colo-rectal, breast, lung, prostate, uterus, bladder, stomach, lymphoma, and leukemia. The 18 centers provide (18x17)/2=153 pairs of codes for each data item on each abstract. The average disagreement over these 153 pairs is summarized in Table 1 (definition of disagreement and other pertinent details are available in the publications cited.)

The message from Table 1 is that variability in coding even under carefully controlled test conditions is high for some data items. The last two items in Table 1 have

Table 1   Major Disagreements for All Center-to-Center Pairwise Comparison of Codes

|  | Average disagreement | Percent of 153 pairs that disagreed on | | |
| --- | --- | --- | --- | --- |
|  | Percent | 0 or 1 codes on 25 abstracts | 2 to 4 codes on 25 abstracts | 5 or more codes on 25 abstracts |
| Histology | 9 | 33 | 60 | 7 |
| Stage | 23 | 0 | 67 | 33 |
| Date of diagnosis | 12 | 20 | 63 | 17 |
| Date of last contact | 10 | 74 | 25 | 1 |

particular relevance to survival analysis, while the results
for histology are a warning signal of the potential
difficulties to be encountered in this classification.
Additional evidence on the magnitude of this important
source of variation in cancer studies is available from the
report of the Beahr's Working Group to Review the Breast
Cancer Detection Demonstration Projects (BCDDP) (Beahrs
et al. (1977)).  Through June 30, 1976 this cooperative
screening endeavor had detected 1810 cancers of which 33%
(592) were classified as minimal.  Of the 592 minimal
cancers 506 were reviewed by a panel of four pathologists
chosen for their particular competence in breast pathology.
The review of the slides was performed independently by the
four pathologists, and then consensus was sought through
joint review where differences of opinion existed.  The
results are shown in Table 2.

The data in Table 2 show several things.  First, some
inference concerning individual input to the consensus can
be made.  Second, 66 cases (15 percent) judged to be cancer
by the project pathologists were by expert consensus judged
to be benign.  Third, non-trivial variation in
classification exists among the independent readings of the
four experts, including 22 cases (4%) where no consensus
could be reached.  Recalling that the pathologist input to a
cancer diagnosis in hospital practice is determining in a
large proportion of cases, it would seem prudent to include
provision for central slide review in planning ad hoc
studies whenever possible.

Table 2   Concordance of Individual and Consensus Diagnosis in Minimal Breast Cancer Cases Diagnosed in BCDDP[1]

| Consensus diagnosis | Pathologist | Individual diagnoses | | | Total |
|---|---|---|---|---|---|
| | | Not ca. | In situ ca. | Infiltrating (<1 cm.) | |
| Not ca. | A | 65 | – | 1 | 66 |
| | B | 49 | 14 | 3 | |
| | C | 61 | 4 | 1 | |
| | D | 65 | 1 | – | |
| In situ ca. | A | 11 | 252 | 7 | 270 |
| | B | 28 | 238 | 4 | |
| | C | 21 | 240 | 9 | |
| | D | 26 | 238 | 6 | |
| Infiltrating (< 1 cm.) | A | 0 | 1 | 103 | 104 |
| | B | 1 | 9 | 94 | |
| | C | 0 | 5 | 99 | |
| | D | 0 | 10 | 94 | |
| Subtotal | | | | | 440 |
| Infiltrating (> 1 cm.) | | | | | 44 |
| No consensus | | | | | 22 |
| Total | | | | | 506 |

1 Adapted from Feigl et al. (1983).

Another obvious weakness of most existing data systems in the cancer area is that the available records do not uniformly include information on exposure variables, e.g. occupational history, smoking history, diet, radiation exposure, residence history. Given the generally long latent period between carcinogenic exposure and clinical manifestations and the geographic mobility of persons in the United States, and given the variable practices of the health professionals responsible for producing the records, it seems unlikely that any significant improvement will occur in this area in the future. This observation has at least two consequences: (1) there will be a continuing need for special epidemiologic studies of cancer etiology utilizing the full armamentarium of study designs, and (2) it will be difficult to utilize directly the available data for direct comparison with the cancer experience of particular groups of individuals, e.g. rubber workers in Dayton, Ohio, or shipyard employees in Seattle, Washington, since many of the exposure variables are of overriding importance.

On the positive side, the growing availability of true incidence data for particular localities in large systems such as SEER will permit surveillance over time that was not possible in the past except for those tumors where diagnosis was uniformly followed by death in a short time period (pancreas cancer and certain forms of lung cancer being good examples). Careful examination of the trends with time and

geography should alert us to the introduction of new
carcinogens into our environment, iatrogenic consequences of
new therapies, and related problems.  Evidence that this is
more than a theoretical possibility comes from surveillance
of the rates of endometrial carcinoma in the San Francisco
Bay area (Austin and Rae (1978)).

In Table 3 the regular increase in incidence in post-
menopausal women in all counties as contrasted with the lack
of temporal pattern in the younger women is striking.  The
threefold increase in wealthy Marin County is particularly
noteworthy.  Clearly, something of concern was happening in

Table 3   Carcinoma of the Corpus Uteri, San Francisco Bay
          Area, 1969-1975 (Annual Incidence Rates
          (per 100,000))

| County | Age | 1969 | 1970 | 1971 | 1972 | 1973 | 1974 | 1975 |
|---|---|---|---|---|---|---|---|---|
| Alameda | 25-49 | 14.3 | 15.7 | 11.0 | 16.8 | 21.9 | 16.2 | 11.8 |
|  | 50-74 | 119.1 | 126.7 | 134.2 | 172.1 | 193.4 | 194.9 | 180.8 |
| Contra Costa | 25-49 | 13.3 | 11.9 | 5.9 | 16.1 | 12.5 | 6.9 | 17.0 |
|  | 50-74 | 132.8 | 122.4 | 152.7 | 172.7 | 186.3 | 210.8 | 256.5 |
| Marin | 25-49 | 11.4 | 11.3 | 12.8 | 8.2 | 29.7 | 16.5 | 8.0 |
|  | 50-74 | 102.4 | 143.8 | 170.3 | 174.1 | 191.3 | 315.1 | 341.1 |
| San Francisco | 25-49 | 8.7 | 16.5 | 17.9 | 17.0 | 15.8 | 16.0 | 21.7 |
|  | 50-74 | 77.3 | 106.6 | 125.4 | 110.0 | 118.2 | 142.9 | 150.0 |
| San Mateo | 25-49 | 13.9 | 9.3 | 13.8 | 10.3 | 11.3 | 9.0 | 12.3 |
|  | 50-74 | 88.8 | 111.8 | 117.8 | 141.5 | 175.0 | 152.6 | 210.0 |

the Bay Area during these years. Subsequent turning down of
these rates in post-menopausal women (not shown), following
reduction in sales of exogenous estrogen preparations used
in the treatment of women undergoing and following
menopause, has led to the interpretation that exogenous
estrogens were the culprit.

Close examination of the accumulating data in the SEER
system is likely to unearth similar episodes in the years
ahead. The continuing increase in female lung cancer rates
is another situation that will be watched closely. The
impression of some gynecologists of increasing numbers of
cases of cervical cancer in younger women is another. While
this surveillance is in one sense comforting, in that our
environments are being thus monitored, it also makes it
clear that historical comparisons or concurrent comparisons
of different groups will be difficult to interpret. The
continuing need for investigations such as case-control
studies, clinical trials, cohort studies, or environmental
modification experiments should be evident.

Finally, it should be noted that population based
registry operations such as SEER and the mortality records
of the U.S. Vital Statistics Systems and its individual
state components, make the prospects for identifying all the
cases of cancer or all deaths from or with cancer in a
cohort or roster of named individuals in a given
geographical area a potentially feasible undertaking.
Computer matching programs have been developed that utilize

a person's first and last name and middle initial, plus
limited identifiers such as sex, date of birth, and social
security number to identify matches on the files. From 1979
deaths onward the National Death Index (U.S. Department of
Health and Human Services (1981)) may be used this way to
identify, e.g. persons in an industrial cohort, or in a
particular union, who have died. Death certificates can be
obtained for the identified deaths. Several of the SEER
sites have the same capability for matching against the
cancer incidence files for their areas. This resource may
prove useful for the types of studies considered in this
book.

REFERENCES


Austin, D.F. and Rae, K.M. (1978). *Uterine Cancer
        Incidence, San Francisco Bay Area*, 1960-1975.
        Resource for Cancer Epidemiology Report, Volume 5,
        Number 1, Department of Health, State of California,
        Berkeley.

Beahrs, O.H., Shapiro, S., and Smart, S. (1977). *Report of
        the Working Group to Review the NCI/ACS Breast
        Cancer Detection Demonstration Projects*. National
        Cancer Institute, Bethesda.

Feigl, P., Polissar, L., Lane, W.W., *et al.* (1983).
        Reliability of basic cancer patient data. *Cancer*,
        (to appear)

Fraumeni, J.R., Jr. (ed.) (1975). *Persons at High Risk of
        Cancer, An Approach to Cancer Etiology and Control*.
        Academic Press, New York.

Fred Hutchinson Cancer Research Center (1981). *Centralized
        Cancer Patient Data System Acquisition Manual,
        Version 2*. Statistics and Quality Control Center,
        Seattle.

Gittlesohn, A. and Senning, J. (1981). Studies of cause of
    death and hospital record diagnoses: I. Comparison
    of cause of death and hospital record diagnosis.
    *American Journal of Public Health*, 69: 680-689.

Ho, J.H.C. (1972). *Nasopharyngeal Carcinoma (NPC), Advances
    in Cancer Research*, 15: 57-92 (G. Klein and
    S. Weinhouse, eds.). Academic Press, New York.

Maclennan, A., Muir, C., Steinitz, R., and Winkler,
    A. (1978). In *Cancer Registration and Its
    Techniques* (W. Davis, ed.). International Agency
    for Research on Cancer, Lyon.

Percy, C., Stanek, E., and Gloeckler, L. (1981). Accuracy
    of cancer death certificates and its effect on
    cancer mortality statistics. *American Journal of
    Public Health*, 71: 242-250.

Schottenfeld, D. and Fraumeni, J.R., Jr. (eds) (1982).
    *Cancer Epidemiology and Prevention*. W.B. Saunders
    Company, Philadelphia.

U.S. Department of Health, Education, and Welfare. (1974).
    *The Health Consequences of Smoking*. U.S. Government
    Printing Office, Washington, D.C.

U.S. Department of Health, Education, and Welfare. (1975).
    *Third National Cancer Survey: Incidence Data, NCI
    Monograph 41*. U.S. Government Printing Office,
    Washington, D.C.

U.S. Department of Health, Education, and Welfare. (1976).
    *Cancer Patient Survival, Report Number 5*. U.S.
    Government Printing Office, Washington, D.C.

U.S. Department of Health and Human Services. (1981).
    *Surveillance, Epidemiology, and End Results:
    Incidence and Mortality Data, 1973-1977, NCI
    Monograph 57*. U.S. Government Printing Office,
    Washington, D.C.

U.S. Department of Health and Human Services. (1981). *User's
    Manual, The National Death Index*. U.S. Government
    Printing Office, Washington, D.C.

Waterhouse, J., Muir, C.S., Comea, P., *et al.* (1976).
    *Cancer Incidence in Five Continents, Vol. III*.
    Scientific Publication Number 15, International
    Agency for Research on Cancer, Lyon.

Weiss, N.S., Szekely, D.R., English, I.R., *et al.* (1979).
    Endometrial cancer in relation to patterns of
    menopausal estrogen use. *Journal of the American
    Medical Association*, 242: 261-264.

2
# STATISTICAL EVALUATION
# OF THE RISK OF CANCER MORTALITY
# AMONG INDUSTRIAL POPULATIONS

Michael J. Symons[*]
University of North Carolina at Chapel Hill
Chapel Hill, North Carolina

John D. Taulbee[†]
The Procter and Gamble Company
Cincinnati, Ohio

## I. INTRODUCTION

Identifying a mortality excess for all cancers or a site-specific cancer in an industrial population is the focus of this chapter. Any approach must account for demographic variables, such as age, sex, and race of the members of the workforce, since the mortality burden of a population typically varies with these characteristics. Other exposure

---

[*]This work was finished while MJS was on leave with an Intergovernmental Personnel Act appointment at the U.S. Environmental Protection Agency, Research Triangle Park, NC.

[†]A major portion of the work was completed while JDT was on the faculty of the Department of Biostatistics and Occupational Health Studies Group, University of North Carolina at Chapel Hill, NC.

variables, most notably smoking, are also of interest as they may contribute to the occurrence of some causes of death. For example, smoking has been thought to be associated with 80% of all lung cancer deaths. The attention of this chapter will be concentrated on statistical methods historically employed for these purposes. However, some epidemiologic perspectives on these approaches cannot be separated. These issues will be considered as they arise.

Two methods have been used most often to identify a mortality excess in an occupational group. The comparison of the proportion of all deaths that are due to the cause of interest in an industrial population with that in some reference group is the most basic. The standardized proportional mortality ratio (SPMR) is the usual summary statistic. The second involves essentially the calculation of an indirect adjusted mortality rate. The standardized mortality ratio (SMR) is the ratio of the observed and expected numbers of deaths. These two approaches are described in the next two sections.

A discussion of these approaches is then presented. This includes concerns about relevant comparison groups, the relationship of the summary statistics from these two methods, and criticisms of these statistics. Alternative procedures to deal with the difficulties that are identified are sketched and referenced. Concern for the overall

significance of multiple tests performed in a single study
is also expressed and a suggestion is provided for
interpreting findings from such studies.  Citations of
studies in the published literature that illustrate these
methods are included throughout the text.


## II.  PROPORTIONAL MORTALITY APPROACH

A proportional mortality study requires the death
certificates of those deceased from a workforce over a
specified period of time.  Using conventions well
established by the World Health Organization in the
International Classification of Diseases, for example the
Eighth Revision Adapted (1968), the attending physician's
description on the death certificate is translated into
numerical codes of the underlying and contributing causes of
death.  From these so-called nosologized death certificates,
the deaths can be tabulated into meaningful categories of
cause of death according to age, race, and sex.  The
comparison is between the number of deaths classified as due
to the cause of interest with that expected, based upon a
comparison population.  The number expected is determined by
applying the proportion of deaths due to the cause category
of interest in the comparison population to the total number
of deaths in the study group for the same category of age,
race, and sex, and then summing over these categories.

A.  Notation and Statistical Fundamentals

For a particular sex-race group, let $d_{ij}$ be the observed
number of deaths in the $i^{th}$ age band and for the $j^{th}$ cause
of death occurring in the study population and let $D_{ij}$ be
the corresponding number in the comparison population.  From
the comparison population the proportion of deaths due to
cause j in age band i is calculated as

$$P_{ij} = D_{ij} / D_{i+} ,  \qquad (1)$$

where $D_{i+}$ is the total number of deaths from all causes in
the $i^{th}$ age band.  The expected number of deaths of cause j
in age group i is calculated by applying this proportion to
the total number of deaths from all causes observed, $d_{i+}$, in
the corresponding age band of the study population, as
follows:

$$e_{ij} = d_{i+}P_{ij} .  \qquad (2)$$

The standardized proportional mortality ratio (SPMR)
then is the ratio of the total observed deaths of cause j
from all age bands and the corresponding total number of
deaths expected.  Visually,

$$\text{SPMR} = \Sigma d_{ij}/\Sigma e_{ij} = d_{+j}/e_{+j} = O/E , \qquad (3)$$

where each summation is over the I age groups.  Consequently
this SPMR is age standardized.  That is, age specific
proportions for the cause of death of interest are applied
to the corresponding observed total deaths in the study
population, giving the expected numbers for each age band.

When the observed number of cause j deaths exceeds (or
is exceeded by) the number expected, the statistical

significance of the excess (or deficit) is of interest. Two
approaches are presented; the first is based upon a Poisson
approximation and the second utilizes the Cochran-Mantel-
Haenszel (CMH) procedure.

First presume a Poisson distribution for the number of
cause j deaths in each age group. Considering applications
for site-specific cancer deaths, this assumption is a
reasonable approximation to the binomial character of the
actual situation. The $P_{ij}$ are typically smaller than 15%,
facilitating this assumption. It follows that the total
number of cause j deaths in all age bands is also Poisson
distributed since the age group sub-totals are readily taken
to be independent of one another. More specifically, we
presume that $d_{ij}$ is distributed as Poisson and that, when
there is no excess or deficit, the mean is $e_{ij}$. The total
number of cause j deaths, $d_{+j}$, then is also Poisson
distributed with mean $e_{+j}$.

The statistical significance of an excess or deficit
can be determined from tables of the Poisson distribution
with parameter $e_{+j}$. Alternatively, when the expected number
of cause j deaths is five or larger, the Poisson
distribution may be approximated by presuming that

$$z = (d_{+j} - e_{+j})/e_{+j}^{1/2} = (0 - E)/E^{1/2} \qquad (4)$$

is normally distributed with mean zero and unit variance.
Transformations to normality for Poisson variables are more
fully discussed in Johnson and Kotz (1969, Chapter 4).

The second approach to examining the statistical
significance of an excess or deficit involves an application

of the CMH procedure, Cochran (1954) and Mantel-Haenszel
(1959). This is preferred to the above Poisson
approximation, especially with small samples. For each of
the age bands, the number of deaths due to cause j and all
other causes in the study and comparison population can be
tabulated in the following two-by-two table:

| Population | Number of cause j deaths | Number of all other deaths | Total deaths |
|---|---|---|---|
| Study | $d_{ij}$ | $d_{i+} - d_{ij}$ | $d_{i+}$ |
| Comparison | $D_{ij}$ | $D_{i+} - D_{ij}$ | $D_{i+}$ |
| Total | $d_{ij} + D_{ij}$ | $T_i - d_{ij} - D_{ij}$ | $T_i$ |

The comparison of the observed number of deaths, $d_{ij}$,
in the study population is made within each of the I age
bands. The $d_{ij}$ is compared with its expectation:

$$E(d_{ij}) = [(d_{ij} + D_{ij})/T_i]d_{i+} = e'_{ij} , \quad (6)$$

where the quantity in square brackets is the combined or
pooled proportion of cause j deaths calculated under the
null hypothesis of no difference in the two groups, Cochran
(1954). The size of the difference, $d_{ij} - e'_{ij}$, must be
scaled by its standard deviation to put it in perspective
relative to its variability with the formula

$$z = (d_{ij} - e'_{ij})/[V(d_{ij})]^{1/2} \quad (6)$$

where $V(d_{ij})$ is the variance of $d_{ij}$.

Mantel and Haenszel (1959) suggest that the variance of
$d_{ij}$ should be computed presuming a hypergeometric model.

Assuming all four margins are fixed, as contrasted with only $d_{i+}$ and $D_{i+}$ fixed in Cochran (1954) where two independent binomials are considered, the variance is

$$V(d_{ij}) = \frac{(d_{i+})(D_{i+})(d_{ij} + D_{ij})(T_i - d_{ij} - D_{ij})}{T_i^2(T_i - 1)} \; . \qquad (7)$$

This allows the CMH procedure to be applied even when the frequencies in a two-by-two table are extremely small. The variance (7) is sometimes conservatively approximated, Peto and Pike (1973), by a Poisson assumption that $V(d_{ij})$ is equal to $e'_{ij}$. When $D_{ij}$ and $D_{i+}$ are very large relative to $d_{ij}$ and $d_{i+}$, this assumption provides essentially the previous procedure. Specifically, the $e'_{ij}$ in equation (6) are then practically the same as the $e_{ij}$ in equation (2).

The CMH procedure utilizes the central limit theorem result that a sum (over the age bands) of the random variables $d_{ij}$ is approximately normal with corresponding mean and variance, specifically

$$E(d_{+j}) = \sum_{i=1}^{I} E(d_{ij}) = \sum_{i=1}^{I} e'_{ij} \qquad (8)$$

and

$$V(d_{+j}) = \sum_{i=1}^{I} V(d_{ij}) \; , \qquad (9)$$

since the deaths in the different age bands are naturally considered as independent. The test statistic is then

$$z' = (\Sigma d_{ij} - \Sigma e'_{ij})/[\Sigma V(d_{ij})]^{1/2} \; , \qquad (10)$$

an approximate standard normal variate, the sums being over the I age bands. The square of this is usually computed and

reported as the one degree of freedom CMH chi-square. It is
most powerful in detecting excesses (deficits) that are
consistently present in all strata, here age bands. See
Peto and Peto (1972). As another minor note, the
controversial Yates continuity correction is not included in
equation (10), but is recommended by Mantel and Haenszel.

A summary statistic, as represented by the SPMR, is not
exactly available from a CMH procedure. Their procedure
provides a sound means of testing the equality of
proportions of cause j deaths in a study and a comparison
population. However, the odds ratio estimate from a CMH
analysis, Mantel and Haenszel (1959), namely,

$$OR = \sum_{i=1}^{I} [d_{ij}(D_{i+} - D_{ij})/T_i] / \sum_{i=1}^{I} [(d_{i+} - d_{ij})D_{ij}/T_i] \quad (11)$$

is a summary statistic commonly reported. It represents an
approximation to the relative risk of a study population
death being of cause j to that of a comparison population
death being of cause j. The approximation is better when
the cause of death is rare; see Cornfield (1951). The test
result for the statistic z' in the standardization (10) is
equivalent to testing the hypothesis that the odds ratio is
unity.

B.   Examples with Discussion

1.   *Sulfur Dioxide and Lung Cancer*
    Consider a study of mortality of male employees of a
chemical plant. A risk of excess lung cancer related to the

production of sulfur dioxide ($SO_2$) was of interest. The
following steps highlight those essential to the process:
(1) Identify all the deaths of workers at the chemical
plant during the period of interest. Table 1 shows these
deaths over the period 1 January 1950 through 31 December
1977. No deaths occurred among these workers in 1976 or
1977.

Identification of all deaths may be a problem. The
completeness of death ascertainment can best be assured by
knowing the vital status at the end of the study period of
all those who might have died during that period. This,
however, is the information required for the study design
described in Section III, typically more expensive to obtain
than collecting only death certificates. If there is a
death benefit available, completeness of the death
ascertainment is enhanced. Cross-checking within company
sources, such as the personnel and retirement departments,
and external sources, for instance union office records, is
helpful to ensure completeness of death ascertainment. Of
course any unknown deaths are bothersome as they may not be
distributed according to causes in the same way as those
identified.

(2) Obtain the cause of death coding, sex, race, age at
death, and other available variables relevant to the cause
of death being studied. For this example, a smoking history
was not available although the number of years worked in the
$SO_2$ production process was available. When latent periods
are of concern, the date of first exposure is needed.

Table 1 Listing of Deaths, 1/1/50-12/31/77, SO$_2$ Production Plant, All Males

| ID of deceased | Cause of death[1] | Date of death | Age at death | Years worked in SO$_2$ plant | Race[2] |
|---|---|---|---|---|---|
| 1 | Acute coronary occlusion | 04/18/67 | 76 | 45 | W |
| 4 | Coronary insufficiency | 09/23/65 | 63 | 40 | B |
| 6 | Cardiovascular | 11/22/75 | 83 | 36 | B |
| 7 | Carcinomatosis | 03/30/61 | 53 | 36 | W |
| 9 | Carcinoma-liver | 11/24/61 | 61 | 35 | B |
| 10 | Coronary occlusion | 12/04/73 | 65 | 31 | B |
| 13 | No record | 06/10/62 | 75 | 29 | W |
| 14 | Acute cardiac failure | 01/11/65 | 67 | 29 | B |
| 15 | No record | 06/12/67 | 81 | 29 | B |
| 16 | No record | 03/30/71 | 76 | 27 | B |
| 18 | Myocardial infarction | 01/22/63 | 66 | 25 | B |
| 19 | Hypertensive cardiovascular | 05/08/72 | 70 | 24 | B |
| 21 | No record | 03/07/53 | 68 | 20 | B |
| 23 | Arteriosclerosis | 04/29/57 | 71 | 19 | W |
| 24 | Nephritis | 11/25/53 | 65 | 17 | B |
| 25 | Cardiovascular | 06/29/68 | 67 | 15 | B |
| 26 | Hypertensive cardiovascular | 02/28/64 | 54 | 12 | B |
| 27 | Myocarditis | 04/25/50 | 72 | 12 | W |
| 28 | Nephritis | 04/29/52 | 41 | 12 | B |
| 29 | Carcinoma-rectum | 06/13/73 | 55 | 9 | W |
| 30 | Carcinoma-lungs | 07/04/67 | 65 | 4 | B |
| 31 | Coronary thrombosis | 08/26/67 | 48 | 3 | W |
| 32 | Acute myocardial infarct | 06/25/75 | 60 | 41 | W |

1. This is the underlying cause of death as abstracted from the death certificate. Cancer of the liver was mentioned as a contributing cause in one death, #28 or #24. These categories can be related to ICDA codes after being nosologized.

2. W = White, B = Black.

Frequently in occupational health applications relevant additional information is not available. For a variable such as smoking which represents another potential contributor to the specific force of mortality of interest, a comparable level of smoking is implicitly presumed present in the comparison population. When a variable such as time from first exposure is available, a method to examine for a trend toward an excess in the cause of death of interest over categories of this variable is sketched in Section II C.

(3) Tabulate the deaths into categories of the various variables of interest. For these deaths, Tables 2 and 3 are sufficient. Table 3 corresponds to the categorization of the professionally nosologized death certificates. The International Classification of Diseases Adapted (ICDA), 8th Revision was considered appropriate given that the study period runs from 1 January 1950 through 31 December 1977. U.S. general population figures for 1968 were used. The choice of the comparison population is discussed more with the next example and in Section III C.

(4) Compute the expected numbers of deaths. In Table 4 the steps for calculating age standardized proportional mortality ratios, separately for blacks and whites, are numerically illustrated. This supplements the algebraic description of SPMRs in Section II A. The age-race standardized proportional mortality ratio is computed as the ratio of the observed number of lung cancer deaths over all

Table 2    Distribution of Death from all Causes by Age and
           Race at Death Among $SO_2$ Plant Employees, 1950-77

| Age at death | White males Number | White males Per cent | Black males Number | Black males Per cent | All males Number | All males Per cent |
|---|---|---|---|---|---|---|
| 40-44 | - | - | 1 | 6.67 | 1 | 4.35 |
| 45-49 | 1 | 12.50 | - | - | 1 | 4.35 |
| 50-54 | 1 | 12.50 | 1 | 6.67 | 2 | 8.70 |
| 55-59 | 1 | 12.50 | - | - | 1 | 4.35 |
| 60-64 | 1 | 12.50 | 2 | 13.33 | 3 | 13.04 |
| 65-69 | - | - | 7 | 46.67 | 7 | 30.43 |
| 70-74 | 2 | 25.00 | 1 | 6.67 | 3 | 13.04 |
| 75-79 | 2 | 25.00 | 1 | 6.67 | 3 | 13.04 |
| 80-84 | - | - | 2 | 13.33 | 2 | 8.70 |
| Total | 8 | 100.00 | 15 | 100.01 | 23 | 100.00 |

Table 3    Classification of Deceased by Cause of Death
           Categories and Race

| Category | White | Black | Total |
|---|---|---|---|
| Circulatory system | 5:(#1,23,27,31,32) | 8:(#4,6,10,14) (#18,19,25,26) | 13 |
| All neoplasms[1] | 2:(#7,29) | 2:(#9,30) | 4 |
| Other[2] | 0:(-) | 2:(#24,28) | 2 |
| No record | 1:(#13) | 3:(#15,16,21) | 4 |
| Total | 8 | 15 | 23 |

1.   There is one cancer of the lung, deceased #30.

2.   Both of these deaths were from nephritis.

Table 4   Calculation of Expected Deaths Due to Respiratory System Neoplasms (ICDA) 160-3, 8th Revision

| Color sex age | 1968 U.S. deaths, all causes [1] | 1968 U.S. deaths, respiratory system neoplasms [1] | Proportion of U.S. deaths, respiratory system neoplasms [1] | Total obs. deaths, all causes | Expected deaths, respiratory system neoplasms | Observed deaths, respiratory system neoplasms | PMR [2] |
|---|---|---|---|---|---|---|---|
| White males: | | | | | | | |
| 40-54 | 113,036 | 7,583 | 0.0671 | 2 | 0.1342 | 0 | 0 |
| 55-64 | 175,169 | 14,778 | 0.0844 | 2 | 0.1688 | 0 | 0 |
| 65-84 | 471,527 | 23,137 | 0.0491 | 4 | 0.1964 | 0 | 0 |
| Total | 759,732 | 45,498 | ------ | 8 | 0.4994 | 0 | 0 |
| Black males: | | | | | | | |
| 40-54 | 23,442 | 1,388 | 0.0592 | 2 | 0.1184 | 0 | 0 |
| 55-64 | 24,704 | 1,638 | 0.0663 | 2 | 0.1326 | 0 | 0 |
| 65-84 | 43,178 | 1,696 | 0.0393 | 11 | 0.4323 | 1 | 2.313 |
| Total | 91,324 | 4,722 | ------ | 15 | 0.6833 | 1 | 1.464[3] |

1. These figures are adapted from the U.S. Department of Health, Education and Welfare publication: Vital Statistics of the United States, 1968, Vol. II Mortality, Section I. General Mortality. Public Health Service, National Center for Health Statistics. U.S. Government Printing Office, Washington, D.C. 1970, pp. 1-40 to 1-205 in part A.

2. PMR is the proportionate mortality ratio or ratio of observed over expected.

3. PMR here is an SPMR, standardized PMR, here standardized for age by calculating expected deaths within each of three age ranges and totalling to get the expected number of deaths for all ages combined, separately for each race.

age-race categories, namely $d_{+j}$ = 1, and the corresponding
expected number, namely $e_{+j}$ = 0.1342 + 0.1688 + 0.1964 +
0.1184 + 0.1326 + 0.4323 = 1.1827 (= 1.2). Notice that the
index i now is used to run over the six age-race categories,
rather than the three age categories. The age-race
standardized PMR is then 1/1.1827 = 0.8455.

Calculations using consecutive revisions of ICDA may
also be desirable when the study is over long calendar
periods of time. As the criteria for coding various causes
of death change with the ICDA revisions, this may be a
concern when the deaths are collected over 20 or more years
as in this study. The procedure described above is
employed, but repeated for each calendar sub-period with an
appropriate comparison population. The ICDA revision
employed to code the death certificates for the study
population deaths is the same as that for the corresponding
comparison population. The sums are then also taken over
the calendar sub-periods and the SPMR is considered
standardized for calendar time in addition to other
variables such as age, race, and sex. For more details, see
Section III E 1.

(5) Calculate the statistical significance of the results.
Clearly in this illustrative example, the observed
experience is well within the range of statistical
fluctuations, as the observed experience is within one death
of that expected.

However for illustrative purposes and to emphasize
another practical concern, consider the four deaths

(individuals #13, #15, #16, and #21 in Table 1), for which no death certificates were available. As an extreme situation, suppose that all were deaths due to lung cancer. What is the statistical significance of the total of five deaths due to lung cancer, one observed plus four presumed? The calculation of the probability of none, one, ..., up to eight such deaths is given in Table 5. The 1.2 expected deaths was calculated based upon the selected comparison population and distribution of all deaths in the study population over the age-race categories. A formal test of the hypothesis that the underlying expectation is 1.2 deaths can be summarized by reporting the P-value. This is the probability of observing five or more deaths when the average is 1.2. From the right most column of Table 5, we see that the P-value is 0.0078. This is highly significant statistically, and the practical importance of obtaining all death certificates should be clear.

When the expected value is greater than five, the normal approximation indicated in the standardization (4) could be used to calculate the P-value. The square of this statistic is the commonly used chi-square test statistic with one degree of freedom.

(6) Summarize the results and qualify according to the thoroughness of case finding and auxiliary data. For Table 5, the P-value of one observed death is about 0.70, so that no excess of lung cancer for the workforce is a reasonable conclusion. However there are two points worthy of mention.

Table 5   Probability of Various Observed Number of Lung
          Cancer Deaths Given the Total Number of Deaths
          Described in Table 1

| Number of deaths observed | Probability[1] | Probability of at least as many deaths[2] |
|:---:|:---:|:---:|
| 0 | 0.3012 | 1.0000 |
| 1 | 0.3614 | 0.6988 |
| 2 | 0.2169 | 0.3374 |
| 3 | 0.0867 | 0.1205 |
| 4 | 0.0260 | 0.0338 |
| 5 | 0.0062 | 0.0078 |
| 6 | 0.0012 | 0.0016 |
| 7 | 0.0002 | 0.0004 |
| 8 | 0.0000 | 0.0002 |

1.  The probability for each number of deaths, x, is
    calculated using the Poisson distribution and
    presuming 1.2 are expected for lung cancer.
    Specifically,

    $$\Pr[X = x \mid 1.2 \text{ expected}] = e^{-1.2} (1.2)^x/x!,$$

    where x! is x factorial (e.g., 5!(= product of
    5,4,3,2, and 1) = 120) and x = 0,1,2,... . These
    values are in most standard text books on
    statistics.

2.  Obtained from the Probability column by summing from
    the bottom to the top, or taking 1.0000 less the
    likelihood of a lesser numbers of deaths.  For
    example, the entry for x = 2 is 1.0000 - 0.3012 -
    0.3614.

First, there was no direct accounting for smoking, except implicitly in the assumption that this group is similar to the 1968 U.S. general population in its smoking patterns. Secondly, there were four deaths for which no death certificate and consequently no cause of death was available. The possibility of one or more lung cancer deaths among the individuals with missing death certificates combined with the uncertainty about the smoking patterns for this workforce provide typical points of concern in evaluating studies involving reconstructed populations.

2. *Cancer Mortality Among Chemists*

   As another example of a proportional mortality study, the paper by Li, Fraumeni, Mantel, and Miller (1969) on cancer mortality among chemists utilizes the CMH procedure to test for statistical significance. The cancer mortality experience of 3,637 male chemists was compared to that of 9,957 U.S. professional men. The expected numbers were computed using the pooled experience as a standard by equation (6), rather than using equation (2) with the professional men as the comparison population. Four age groups were identified as strata for this analysis.

   In comparing two study groups, the CMH method admits that the rates may be variable in each group, rather than assuming a fixed rate for the comparison population. However, recall that when the comparison population is very large relative to the study population, there will be little to distinguish between the two approaches. The comparison

of chemists with U.S. professional men seems more
appropriate than the comparison of a plant workforce with
the U.S. general population because of a similar screening
for eligibility into professional ranks. Further, the
ability to work is not required of all those in the U.S.
general population as it is of the $SO_2$ plant workforce. The
potential bias that may result is referred to as the healthy
worker effect, about which more is said in Section III C.

The study concluded that about one half of the 90
excess cancer deaths were attributed to the excesses of
malignant lymphomas and carcinomas of the pancreas. The
authors naturally concluded from these findings and other
fragmentary data that chemical agents may induce cancers of
the pancreas and lymphoid in man.

C. Heterogeneous Strata: Variation with Degree of Exposure

The statistical power of the procedures described and
illustrated in the previous two sections rests upon a
consistent excess, or deficit, in each of the strata. Of
fundamental interest may be the lack of consistency over
categories for a variable. For example, a gradual increase
in an age standardized PMR over categories of increasing
years of exposure offers evidence internal to the study that
supports a causal relationship between an exposure and a
cancer excess. Because selection into the workforce over
time may be reasonably similar for all members of the study
population, for example eligibility to work is a minimum

requirement, such an analysis is less suspect than a comparison with a U.S. general population.

A full treatment requires weighted least squares, for example Grizzle, Starmer, and Koch (1969), or survivorship methods, both of which are beyond the practical limits of this chapter. Survivorship methods will be indicated in Section IV D and the simplest weighted least squares approach sketched now.

The special case of two groups of exposure is presented. Let the indices i and j be as before, indexing age groups and causes of death, respectively, but also include an index k = 1 (or 2) to indicate an extensive (or brief) work history of exposure to the agent being examined. The possibility of a greater proportion of cause j deaths among the more extensively exposed individuals can be examined by a straightforward application of the CMH procedure described in Section II A. For each of the I strata construct the table:

| Exposure group | Cause j deaths | All other deaths | Total |
|---|---|---|---|
| Extensive: k = 1 | $d_{ij1}$ | $d_{i+1} - d_{ij1}$ | $d_{i+1}$ |
| Brief: k = 2 | $d_{ij2}$ | $d_{i+2} - d_{ij2}$ | $d_{i+2}$ |
| Total | $d_{ij+}$ | $d_{i++} - d_{ij+}$ | $d_{i++}$ |

(Recall that the presence of a plus sign (+) in the position of an index indicates a summation over all levels of that

index.)  The previously described CMH calculations are
subsequently performed with the tables corresponding to
i = 1, 2, ..., and I, being the strata.  The method is
applicable even when the frequencies in the above table are
small, but of course the sensitivity of the CMH test will be
diminished.

The procedure is preferred to examining a difference in
SPMRs calculated for each of the two groups of exposure.
Even though the same general population standard would be
utilized, differences in the SPMRs could be attributed to
the differences in age distribution of deaths from all
causes over the I strata between exposure groups.  This same
point will be considered further in Section IV A.

III.  HISTORIC PROSPECTIVE COHORT MORTALITY STUDY

A fundamental limitation of proportional mortality studies
is the absence of an evaluation of the total mortality
burden.  It provides information only on the different,
relative cause specific forces of mortality.  Since
exposures often manifest themselves in particular causes of
death, as finding mesotheliomas among asbestos workers, this
still allows a straightforward, quickly completed, and
usually reliable assessment of cause-specific relative
mortality in a working population.  Nevertheless, the
previous analysis involved only the events, here deaths.
The mortality rates are the fundamental objects of interest
in these analyses; the denominator is not part of the

previous analyses, only the numerator. The concern in this section is for an observed number of cause j deaths from among the number of individuals at risk of dying, rather than relative to all those dying.

The cornerstone for a historic prospective cohort mortality study lies in determining the amount of risk of death for the workforce over an observation period. This evaluation requires the combined information on the number of cohort members and period of observation for each member. The term "historic prospective" derives from the identification of all those workers defined to be members of the cohort as of the same past date and followed to a more recent past date. For this period of observation the expected deaths for all causes or a specific cause are calculated. The ratio of the observed deaths to the expected, possibly multiplied by 100 and reported as a percentage, is termed the standardized mortality ratio. The standardization refers to the specific death rates used from a comparison population. For example, the use of white male age-specific death rates for leukemias in the United States, as applied to a white male cohort, would provide an age standardized mortality ratio.

As the basic notation and calculations required have been identified in Section II A, we proceed directly with an example. However, a subscript for the cause of death will no longer be used. A specific cause or all causes may now be the focus for analysis.

A. Essential Features and an Example

A discussion of the experience of a small chemical company
in the northeastern part of the United States should
crystallize the ideas. This is adapted from the
documentation of Pagnotto, Elkins, and Brugsch (1979). A
variety of solvents were utilized in the operations of this
plant, beginning early in 1961. Some of these solvents were
known to have contained benzene and other fractions
suspected or known as carcinogens, at least as contaminants
in the solvents.

A total of 38 male employees worked with these
processes at the outset. Table 6 gives the age distribution
of these workers, an unknown racial mixture of primarily
whites and blacks. Follow-up of these workmen through the
end of the observation period, 31 December 1977, was
completed. Although a number had left the company before
1977, various follow-up efforts determined that all were
alive at the end of 1977, except for three deaths. The

Table 6   Age Distribution in Males Employed on 1 January
          1961 in Operations Involving Solvents

| Age in 1961 | 21 | 22 | 25 | 26 | 27 | 28 | 29 | 30 | 33 | 35 | 36 |
|---|---|---|---|---|---|---|---|---|---|---|---|
| Number of male employees | 3 | 1 | 4 | 2 | 1 | 1 | 2 | 1 | 2 | 3 | 1 |
| Age in 1961 | 38 | 39 | 42 | 43 | 49 | 51 | 53 | 54 | 57 | 61 | 62 | 65 |
| Number of male employees | 2 | 1 | 2 | 1 | 2 | 1 | 2 | 1 | 2 | 1 | 1 | 1 |

Table 7   Description of the Three Deaths Among Cohort
          Presented in Table 1

| Age in 1961 | Year of death |
|-------------|---------------|
| 53 | 1972 |
| 57 | 1975 |
| 61 | 1972 |

years of death and the 1961 ages of these three workmen are
shown in Table 7.  Inspection of their death certificates
revealed that none of these deaths was related in any way to
cancer.  This observation was put forward as suggesting that
there was no increased risk of dying from leukemia for
individuals working in these operations.  As the exposures
involved were shown to sometimes exceed the existing
standard for benzene, it was also concluded that there was
no evidence to support a reduction in the benzene standard.

Additional analysis is needed in order to fully examine
this claim.  In particular, one would like to know how many
deaths due to leukemia and related cancers (ICDA codes
200-209) would be expected among these workers over this
period of time.  This is accomplished by tabulating the
person-years of experience and applying age-specific death
rates for the causes of interest from a comparison
population.  The all males age-specific death rates
(combined over races as this separation was not available)
are desired from a population with comparable basic biologic
risk to leukemia as a cause of death.  If this population is

similar, except for the solvent exposures from working in the industry, then an excess of observed to expected could be attributable to the work environment exposures.

Before proceeding to the concept and calculation of person-years, the essential features of this study can be schematically presented as in Figure 1. A cohort of workers, each of whose eligibility was defined as of a date in the past, is followed and vital status determined as of the end of the observation period. The shaded areas represent incompleteness in the enumeration of all possible workers (B), unknown deaths at the end of the study (D), and unknown cause of death for a decedent (E). The percentage of

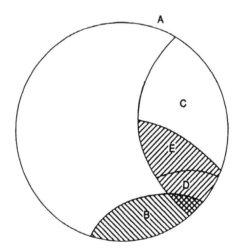

Figure 1      Schematic of Essential Features of Complete
              Historic Prospective Cohort Study.  A. All
              eligible workers as of a date in time.  B. Unknown
              members of the cohort defined by A after enumer-
              ation.  C. All deaths among cohort members.
              D. Unknown deaths after vital status check at end of
              observation period.  E. Deaths of which no death
              certificate is available.

certificates obtained for those known deaths, the proportion
lost to follow-up (that is the fraction of individuals
identified at the outset but with unknown vital status at
the end of the study), and proportion of the cohort
enumerated should be reported as part of the study results.
These proportions bear directly on the reliability of the
conclusions derived, since they describe amounts of missing
data in the study. The proportion of death certificates
available for those known decedents is directly calculable.
The proportion enumerated that are lost to follow-up is
directly calculable and is an upper bound on the number of
unknown deaths. The proportion of the defined cohort that
is enumerated is not really directly estimable, but
describing the diligence of cross checking various sources,
internal to the industry and external ones (for example
local union and state and/or federal government offices), to
ensure completeness of the enumeration of all cohort members
enhances the study validity.

The exact methods to be employed would be expected to
vary with each situation. Andjelkovic, Taulbee, and Symons
(1976) and Pell (1978) provide illustration for specific
situations that can be adapted. Marsh and Enterline (1979)
describe a method of verification of the enumeration based
upon IRS records which is completely independent of company
held records and is recommended. Lloyd and Ciocco (1969)
give a detailed description of the follow-up methods they
employed in their study of mortality among steel workers.

These procedures, supplemented with some ingenuity for a
specific situation, should provide adequate enumeration of a
workforce and subsequent follow-up.


B.  Calculation of Person-Years

The concept of a person-year is simply that of one person
being alive, or exposed to the chance of dying, for one
year; see Hill (1972).  The total experience of the cohort
of workers over the observation period is obtained by
determining the experience of each worker during the
observation period and summing these contributions over the
members of the cohort.

The idea is illustrated by an approximate technique for
the example of this section.  In order to accumulate the
amount of worker experience, it is necessary to age the
individuals in Table 6 over the 17 year observation period.
The number of years of experience, or exposure to the risk
of dying, for example by five-year age groups, is needed as
an intermediate step in the calculation of an age
standardized mortality ratio.  This is determined by aging
each individual from his year of age in 1961 through single
years of calendar time over the study period.

The complete details are given in Table 8.  Attention
to one or two individuals will illustrate the process.  For
example, there was one individual 22 years of age on 1
January 1961.  He was alive on 31 December 1977 so a tally
mark was made in each appropriate box, single year of age by

Table 8    Tabulation and Calculation of Person-Years of Experience by Calender Year and Age of the Cohort Members Identified in Table 1 with Tally Marks for Individual by Single Years of Age and Calender Time on the Left

| Age | Calendar year 61 | 66 | 71 | 76 | 77 | Single years of age Col (1) | Person-years of experience by 5-Year totals: Birthdate on December 31 Col (2) | 5-Year totals: Birthdate on January 1 Col (3) | 5-Year totals: Average of two other columns $n_i$ Col (4) |
|---|---|---|---|---|---|---|---|---|---|
| 20 | | | | | | 0 | | | |
| | | | | | | 3 | | | |
| | | | | | | 4 | 15 | 11 | 13.00 |
| | | | | | | 4 | | | |
| | | | | | | 4 | | | |
| 25 | | | | | | 8 | | | |
| | | | | | | 10 | | | |
| | | | | | | 11 | 55 | 45 | 50.00 |
| | | | | | | 12 | | | |
| | | | | | | 14 | | | |
| 30 | | | | | | 15 | | | |
| | | | | | | 15 | | | |
| | | | | | | 15 | 79 | 76 | 77.50 |
| | | | | | | 17 | | | |
| | | | | | | 17 | | | |
| 35 | | | | | | 20 | | | |
| | | | | | | 21 | | | |
| | | | | | | 21 | 102 | 99 | 100.50 |
| | | | | | | 20 | | | |
| | | | | | | 20 | | | |
| 40 | | | | | | 20 | | | |
| | | | | | | 20 | | | |
| | | | | | | 18 | 91 | 95 | 93.00 |
| | | | | | | 17 | | | |
| | | | | | | 16 | | | |
| 45 | | | | | | 15 | | | |
| | | | | | | 13 | | | |
| | | | | | | 12 | 66 | 68 | 67.00 |
| | | | | | | 12 | | | |
| | | | | | | 14 | | | |
| 50 | | | | | | 12 | | | |
| | | | | | | 13 | | | |
| | | | | | | 10 | 58 | 60 | 59.00 |
| | | | | | | 11 | | | |
| | | | | | | 12 | | | |
| 55 | | | | | | 10 | | | |
| | | | | | | 9 | | | |
| | | | | | | 11 | 50 | 53 | 51.50 |
| | | | | | | 11 | | | |
| | | | | | | 9 | | | |
| 60 | | | | | | 8 | | | |
| | | | | | | 9 | | | |
| | | | | | | 10 | 46.5 | 46.0 | 46.25 |
| | | | | | | 10 | | | |
| | | | | | | 9.5 | | | |
| 65 | | | | | | 10 | | | |
| | | | | | | 8 | | | |
| | | | | | | 8 | 40 | 42.5 | 41.25 |
| | | | | | | 7 | | | |
| | | | | | | 7 | | | |
| 70 | | | | | | 6 | | | |
| | | | | | | 4.5 | | | |
| | | | | | | 3.5 | 19 | 24 | 21.50 |
| | | | | | | 3 | | | |
| | | | | | | 2 | | | |
| 75 | | | | | | 2 | | | |
| | | | | | | 2 | | | |
| | | | | | | 2 | 9 | 10 | 9.50 |
| | | | | | | 2 | | | |
| | | | | | | 1 | | | |
| 80 | | | | | | 1 | | | |
| | | | | | | 1 | 2 | 3 | 2.5 |

single year of calendar time.  This individual then
contributed one person-year of experience to age 22 in 1961,
to age 23 in 1962, ..., and finally contributed one person-
year to age 38 in 1977.  Those dying, or otherwise leaving
the study, contribute on average one-half year of experience
in the calendar year that they exit.  The number of person-
years of experience for each single year of age is given in
the column labelled Col (1) in Table 8, obtained by
totalling over the calendar years of the study period for
that age.  Each five-year age total in the column labelled
Col (2) is the sum of the corresponding five, single year of
age, person-years totals given in column Col (1).

Implicit in this approach is the assumption that the
birthdate is at the end of the year.  A full person-year of
experience is credited to the age-calendar year box by the
age at the beginning of the year.  The birthdate would be
more accurately assumed to be at mid-year, on average.  To
correct this bias of more person-years in the younger age
groups, a second column of person-years was constructed for
each five year age group.  This column is labelled Col (3)
in Table 8.  Col (2) can be thought of as corresponding to
an assumption that the birthdate occurs on 31 December of
each single year.  Col (3) can be thought of as
corresponding to an assumption that the birthdate occurs on
1 January of each year and is obtained by taking the single
year age totals in Col (1) and shifting each to the next age
interval.  For example the five year person-years total in

Col (2) for the age group 35-39 is calculated as the sum:
20 + 21 + 21 + 20 + 20 = 102. The corresponding five year
person-years total in Col (3) is obtained by summing the
single age totals in Col (1) for the ages 34 through 38,
visually, 17 + 20 + 21 + 21 + 20 = 99. As this shifts the
person-years of experience too far toward the older ages,
another column of five year age group totals is added,
labelled Col (4),and computed as the arithmetic mean of the
Col (2) and Col (3).

The notation $n_i$ is used to denote the person-years
experience accumulated in the $i^{th}$ five-year age interval.

This procedure and variations are clearly approximate.
The purpose is primarily to illustrate the ideas of
accumulating person-years of experience, but represents the
basis for some computer programs, e.g., Monson (1974).
Exact accumulations of units of experience in the various
age-calendar years can be accomplished by utilizing exact
dates of birth, death, and exit; see Hill (1972) and Marsh
(1980). For very large data sets, a life table scheme could
be adapted for hand calculations to age each single year or
five-year cohort through the study period.

The person-years of experience for the five-year age
group totals are analogous to a stationary population
distribution in a current life table; see Shyrock and Siegel
(1973, p. 433). In this application the person-years of
experience in any age group can be imagined as a larger
fictitious workforce, exposed to dying at the mid-point of

the study period. From this artificial distribution of
workers, the expected number of deaths from one specific
cause, or all causes, of death is calculated in exactly the
same manner as the expected numbers for the indirect
standardization of mortality rates, as described by Hill
(1961, p. 231). Specifically, the schedule of age specific
death rates, denoting the $i^{th}$ five year specific rate by $R_i$,
from a comparison (standard) population are applied to this
fictitious workforce to determine the number of deaths
expected. In symbols we have

$$e_i = R_i n_i \ , \qquad\qquad (12)$$

representing the calculation of the expected number of
deaths in the $i^{th}$ age group.


C.  Choice of a Comparison Population and the Healthy Worker
    Effect

The choice of a population whose mortality rates will be
applied to the distribution of person-years to obtain the
expected numbers of deaths is simple in concept but
difficult in practice. Conceptually one wishes to apply the
mortality experience of a population selected in a manner
similar to the study population, for example, employed in a
similar industry but without the exposures of the study
population. Further, one would like the comparison group to
be of similar age, race, sex, and socio-economic
composition, or simply to have similar characteristics,
except for exposures, thought to bear on the risk of death

from the cause under study. The difficulty comes in the
uniqueness of most industrial-geographic combinations. As
an industry usually attracts its workforce from the local
geographic population, variations in its social-ethnic
composition are of concern. Often populations that would
make suitable standards for comparisons do not exist or the
desired mortality rates are simply not available.

The U.S. general population is a frequent choice for a
comparison population, the required death rates derived from
U.S. Census and Vital Statistics publications. The
difficulty is that a priori one expects an employed
population to be healthier and consequently bear a lighter
overall mortality burden by virtue of its selection and
maintenance as an employed group. The U.S. general
population includes those unable to work. This bias is
referred to as the "healthy worker effect" and is well
known. See for example Hill (1959, p. 231), Andjelkovic *et
al.* (1976), and Monson and Nakano (1976).

A major difficulty in addressing the healthy worker
effect is a lack of a clear definition of the concept. The
necessarily healthier state of an individual presented at
the time of hire is a portion of the idea. There may then
be a "regression to the mean" from this healthier state, an
erosion of health due to deleterious exposures at the work
place, or some combination of these two effects. McMichael
(1976) implies that the healthy worker effect varies with
age, race, and cause of death under study from comparisons

of various occupational groups with U.S. general
populations. The confounded effect of initially healthier
workers exposed to presumably varying exposures at the
workplace is not addressed. Hopefully these effects will be
separable, at least to some degree, as more data become
available on diverse working populations, with ranges of
types and degrees of exposures. For additional comment, see
Section IV C.


D.  Calculation and Statistical Significance of the
    Standardized Mortality Ratio

Table 9 contains the calculations of the expected number of
deaths for leukemia plus related cancers (ICDA: 200-209).
For these data, the 1970 U.S. Census and deaths are used to
calculate standard age-specific rates, $R_i$, for this cause of
death category. These figures are for all races, implicitly
assuming that the racial mix of this workforce is similar to
that of the U.S. The expected number of deaths are obtained
by applying the 1970 U.S. age-specific death rates, $R_i$, to
the person-years of experience, $n_i$, in the corresponding
five-year age-group as shown by equation (12).

    One can see from Table 9 that the total (over all age
groups) number of expected deaths due to leukemia and
related cancers is far less than one (0.16) for this
workforce over the 17 year follow-up. No deaths related to
any type of cancer were observed. Put another way, this
workforce would have had to have been larger by a factor of

Table 9 Calculations of Expected Number of Deaths for Leukemias

| i | age group | Five-year Person-years[1] experience $n_i$ | 1970 U.S.[2] population $N_i$ | 1970 U.S.[3] deaths $D_i$ | Rate $R_i = D_i/N_i$ | Leukemias (ICDA: 200-209) Expected number $e_i = R_i n_i$ |
|---|---|---|---|---|---|---|
| 1 | 20-24 | 13.00 | 7,917,269 | 424 | 0.000054 | 0.000702 |
| 2 | 25-29 | 50.00 | 6,621,567 | 373 | 0.000056 | 0.002800 |
| 3 | 30-34 | 77.50 | 5,595,790 | 339 | 0.000061 | 0.004728 |
| 4 | 35-39 | 100.50 | 5,412,423 | 429 | 0.000079 | 0.007940 |
| 5 | 40-44 | 93.00 | 5,818,813 | 583 | 0.000100 | 0.009300 |
| 6 | 45-49 | 67.00 | 5,851,334 | 955 | 0.000163 | 0.010921 |
| 7 | 50-54 | 59.00 | 5,347,916 | 1216 | 0.000227 | 0.013393 |
| 8 | 55-59 | 51.50 | 4,765,821 | 1685 | 0.000354 | 0.018231 |
| 9 | 60-64 | 46.25 | 4,026,972 | 2053 | 0.000510 | 0.023588 |
| 10 | 65-69 | 41.25 | 3,122,084 | 2281 | 0.000731 | 0.030154 |
| 11 | 70-74 | 21.50 | 2,315,000 | 2371 | 0.001024 | 0.022016 |
| 12 | 75-79 | 9.50 | 1,560,661 | 2098 | 0.001344 | 0.012768 |
| 13 | 80-84 | 2.50 | 875,584 | 1369 | 0.001564 | 0.003910 |
| Total | 20-84 | 632.5 | ---- | ---- | ---- | 0.160451 |

1. Person-years are from Table 8.

2. These figures are from 1970 U.S. Census of Population, U.S. Summary. General Population Characteristics.

3. These figures are from 1970 Vital Statistics of the United States, Volume II. Mortality, Part B.

about six before even one leukemia type cancer death would have been expected. For this study the observed and expected experience, presuming no excesses due to the solvent exposure(s), are in agreement.

Comment is now made on the significance of these data to the benzene standard. Even though the environmental data for benzene exposures were in the neighborhood of the benzene standard and sometimes exceeded it, noting that the latency for leukemia could be greater than the 17 year follow-up period of this study, these data do not seem to address the issue of the risk to leukemia. Consequently a conclusion suggesting a reduction or increase in the benzene standard seems inappropriate.

The standardized mortality ratio (SMR) is simply the ratio of the observed, $0 = \Sigma d_i$, to the expected, $E = \Sigma e_i$, number of deaths. Its limitations are discussed later and alternatives described. However, its simplicity and general relevance account for its almost exclusive use as a summary statistic in occupational mortality studies. In this example the SMR is zero since no deaths for these cancers were observed. In general its significance can be judged by reference to Poisson tables for expectations less than five and by the chi-square approximation $\chi^2 = (0-E)^2/E$, with one degree of freedom, for $E$ greater than or equal to five. The discussion in Section II A is directly appropriate, with one reservation presented next.

The CMH statistic could be adapted for application here as with the proportional mortality study. However, strictly

speaking we have a mid-period approximated workforce as the
standard or comparison group, usually a U.S. general
population. Such data are not identifiable as those dying
during a calendar year or surviving the year, as presented
by Wong (1977) or as needed for the calculation of person-
years in Section III B. As there are serious limitations in
the SMR, we do not feel that further refinements in its
calculation or testing of its significance are warranted.
However, when two cohorts are being compared the CMH
procedure may be appropriate and this application is
sketched at the end of Section IV D.

An approximate interval estimate can be constructed
with the SMR. Presuming the Poisson model is relevant, the
SMR = $0/E$ and $V(0/E) = E(0)/E^2$. Consequently the estimated
standard error of estimate is $(0/E^2)^{1/2}$ since one estimates
$E(0)$ with $0$ when the null hypothesis, $E(0) = E$, may not be
relevant. From central limit theory, appropriate when the
expected number of events is at least five, a $(1 - \alpha)$ 100%
confidence interval is given by

$$0/E \pm Z_{1-\alpha/2} \ (0/E^2)^{1/2} \tag{13}$$

where $Z_{1-\alpha/2}$ is the percentile of the standard normal
distribution with area $\alpha/2$ to its right. This confidence
interval is for the constant multiple $\theta$ of the standard
population specific rates described in Section IV C; see
especially equation (24) and supporting text.

Finally we note that as with the SPMR, the standardized
portion of an SMR derives from the use of specific rates
within the categories of variables relevant to survival, in

this example five-year age groups. Calculations separately
for sex and race, in addition to age, are usually performed,
but summing is usually only over age. This would give age-
standardized mortality ratios for each sex-race group.


E. Further Comments and Variants with the Standardized
   Mortality Ratio

Once the computations of the expected numbers of deaths and
basic ideas of person-years are understood, several
variations can be easily described. Four variations are
presented in this section. In addition, a relationship
between SPMRs and SMRs and some comments on sample size are
included.

1. *Standardization for Calendar Time of Follow-up*
    Concern might be expressed about the summing of the
person-years of experience over the full 17 year observation
period. A more refined calculation of the expected numbers
of deaths may be warranted if there is a trend over calendar
time. Rates for lung cancers have steadily increased from
1930 to 1970. Those for stomach cancers have steadily
decreased. Leukemia rates have risen slightly from 1930 to
1970 but have been virtually constant since 1955. It is a
straightforward matter to adjust for such trends that
overlap with the follow-up period of the study, and may be
of potential concern.
    The procedure would be to repeat the operations
described in Table 8 and 9 for each sub-interval of calendar

time. For example, with the calendar period 1961 through
1977, the sub-periods of 1961-1965, 1966-1970, and 1971-1977
would probably be adequate to represent the effects of most
chronological mortality trends. Standard population rates
at the midpoint of each calendar sub-period would be applied
to the person-years of experience for that calendar sub-
period to determine the corresponding expected numbers of
deaths. These expected numbers would then be totalled
across the calendar sub-periods and a ratio formed of the
observed deaths over this expected number. The resulting
SMR would be age and calendar period standardized for all
males in the leukemia example. See Monson and Nakano (1976)
for an example and Monson (1974) for a computer program.
Shorter intervals on the chronology of follow-up time,
yearly if necessary, are only a computational complication
and do not pose a conceptual problem.

2. *Trends with Years Worked or Levels of Exposure*

The calculation of an SMR for each of several groups of
workers, separated by increasing numbers of years worked or
levels of exposure, may prove informative. As levels of
exposure are rarely available, suppose age-sex-race-calendar
time specific mortality rates are applied separately to
workers with less than 10, 10 through 19, 20 through 29, and
30 or more years of experience. The results may suggest a
trend toward larger SMRs with increasing years of
employment. Caution should be exercised, however, as
indirect rates are not strictly comparable, even when the

same standard rates are used.  This is because the
distribution of those at risk over the categories
corresponding to the specific rates utilized may differ from
one category of years worked to another.  This limitation is
discussed further in Section IV A.  Alternative approaches
are sketched in Sections IV C and especially IV D.

Any fitting of trends in SMRs so formed should be done
with a weighting inverse to the variance of each SMR.  The
estimated variance is given with equation (13), appropriate
when a Poisson model is presumed.

3.  *Modification for Cancer Latency*

To accommodate a lag period for the development of
cancer, the person-years tabulation for each individual is
commenced after a specified number of years from first
exposure to the agent of interest.  This may be repeated for
a sequence of number of years, say 10, 15, 20, 25, and 30
years, from the date of entry to the industry, known date of
introduction of the agent into the plant, or date of first
job working with the agent.  This date and age of the
individual at that time are, of course, needed.

The application of rates for a comparison population to
this distribution of person-years of experience will result
in larger SMRs, if the deaths for the specific cancers do in
fact appear predominantly after a latent period.  The reason
that the SMRs will be larger is that with few deaths from
the cause of interest before the latent period has passed,
there would be a dilution of the excess of deaths occurring

later if this early period of experience and more sparse
death experience was included.

## 4. *Relation of SMRs and SPMRs*

Although the proportional mortality ratios involve only
data on deaths, the standardized proportional mortality
ratio (SPMR) for a particular cause of death empirically and
theoretically turns out to be a fairly good estimator of the
adjusted standardized mortality ratio (SMR) for the same
cause of death. The adjustment to the cause specific SMR is
to divide it by the all causes SMR, making it relative to
the overall force of mortality as the SPMR is by definition.

This is perhaps best illustrated. Table 10 contains
adjusted SMRs and SPMRs for a series of ICDA code groupings.
This is for the same rubber manufacturing plants described
by Andjelkovic *et al.* (1976). The comparability of these
two statistics is quite impressive and the utility of a
proportional mortality analysis should not be overlooked, as
Milham (1978) has so convincingly encouraged.

There is also theoretical support for the relationship.
Kupper *et al.* (1978) describe the nature of the mathematical
relationship of these two statistics, as a parallel linear
combination form of each is exhibited. Unusual conditions
where the correspondence is not so good are presented. In
addition, Breslow (1976) notes that ratios of SPMRs for
different causes estimate the same quantity as ratios of
SMRs for those causes, under specified conditions.

Table 10   A Comparison[1] of the Adjusted[2] SMR with the SPMR for Grouped Causes of Death Among White Male Hourly Workers 40-84 Years of Age, Akron, 1964-73

| ICDA code | Observed deaths | Expected deaths (adjusted SMR) | Expected deaths (SPMR) | Adjusted SMR | SPMR |
|---|---|---|---|---|---|
| 000-136 | 13 | 16.05 | 15.94 | 81 | 82 |
| 140-149 | 7 | 13.21 | 13.40 | 53 | 52 |
| 150-159 | 157 | 136.52 | 134.31 | 115 | 117 |
| 160-163 | 123 | 144.70 | 140.37 | 85 | 88 |
| 170-174 | 7 | 8.43 | 9.21 | 83 | 76 |
| 180-189 | 89 | 74.17 | 74.34 | 120 | 120 |
| 190-199 | 34 | 37.36 | 35.64 | 91 | 95 |
| 200-209 | 55 | 41.67 | 39.73 | 132 | 138 |
| 210-239 | 7 | 5.51 | 4.34 | 127 | 161 |
| 240-279 | 56 | 48.70 | 45.96 | 115 | 122 |
| 280-289 | 7 | 4.07 | 4.68 | 172 | 150 |
| 290-389 | 22 | 23.16 | 22.08 | 95 | 100 |
| 393-398 | 27 | 17.76 | 16.26 | 152 | 166 |
| 400-404 | 29 | 26.61 | 26.24 | 109 | 111 |
| 410-413 | 1023 | 1012.87 | 1027.78 | 101 | 100 |
| 430-438 | 252 | 231.19 | 235.40 | 109 | 107 |
| 470-486 | 46 | 80.70 | 77.45 | 57 | 59 |
| 490-493 | 75 | 78.95 | 79.03 | 95 | 95 |
| 500-519 | 23 | 24.21 | 26.02 | 95 | 88 |
| 520-577 | 78 | 92.86 | 93.12 | 84 | 84 |
| 680-709 | 1 | 1.47 | 2.39 | 68 | 42 |
| 710-738 | 5 | 4.76 | 4.03 | 105 | 124 |
| 780-796 | 28 | 20.29 | 20.36 | 138 | 138 |
| 800-999 | 94 | 108.05 | 101.80 | 87 | 92 |
| All other | 167 | 171.73 | 175.12 | 97 | 95 |

1.   Unpublished work by D. Andjelkovich and J. Taulbee of the Occupational Health Studies Group, 1975.
2.   Adjusted SMR is the cause specific SMR divided by the all causes SMR.

DeCoufle, Thomas, and Pickle (1980) review and also discuss
this relationship.

## 5. *Use of Internal Standards*

The criticism of the healthy worker effect in SMR
calculations can be overcome in certain situations. Lloyd
*et al.* (1970) used the combined mortality of all steel
workers to examine mortality risk in specific work areas.
Andjelkovich *et al.* (1977) used this approach to compare the
cause specific mortality in various portions of a rubber
manufacturing plant. The total plant rates were computed
and used as the (internal) standard to compute the expected
number of deaths in major portions of the manufacturing
process. By not using an external standard such as the U.S.
general population, the selection of the workers for any
specific process is more comparable to that in other
portions of the plants. After adjustment for age and race
differences in these male employees, the comparison of
processes by SMRs should be more valid. Differences in the
distribution of experience, or in the distribution of
deaths, by race and age as discussed in Section IV A,
however, must still be considered.

An analogue to the SMR based on an internal standard is
possible with proportional mortality. It uses the
proportion of all deaths due to a specific cause of death
for all workers of a plant to calculate the number of deaths
expected from this cause in a portion of the total
workforce. Redmond and Breslin (1975) compare the results

with this approach to those with the SMR and an internal
standard and they note that the SPMR with an internal
standard may detect only the more pronounced deviations from
expected. We noted above the good agreement between SMRs
and SPMRs with _external_ standards. As the stability of the
proportion of deaths due to a specific cause of death is
limited by the total number of deaths in the entire
workforce, the finding reported by Redmond and Breslin
(1975) may be due to a corresponding increase in variance.

6. *Comments on Sample Size*

Consider the problem of sample size for an occupational
cohort mortality study. First we note that sample size
refers to the number of person-years, the $n_i$ of Table 8, in
each of the age groups. These can be increased by
enumerating a larger cohort or following a given cohort for
a longer period of time. The latter may be the only way to
increase the number of person-years in each age group, since
most plant sizes are fixed. There are practical limits to
this strategy, or course, as the reported cause of death
after age 75 or 80 becomes less reliable and losses to
follow-up due to other causes of death will deplete the
cohort.

Nevertheless, suppose that an observed SMR of a
multiple $\theta$ or greater would be of interest. The ratio of
observed ($O$) to expected ($E$) is significantly greater than
unity when

$$\sum_{k=0}^{\infty} [\exp(-E) \; E^k/k!] \leq \alpha \; , \tag{14}$$

or since $O/E=\theta$ under an alternative hypothesis, the ratio would equivalently be significantly greater than unity when

$$\sum_{k=0}^{O-1} [\exp(-O/\theta) \; (O/\theta)^k/k!] \geq 1 - \alpha, \tag{15}$$

using the Poisson distribution as illustrated by Table 5. With rates $R_i$ from a standard population, the expected number is given by

$$\sum_{i=1}^{I} R_i n_i = E = O/\theta. \tag{16}$$

One can then see that the SMR is really the constant multiple of age specific rates which generates the observed number of deaths as follows:

$$\sum_{i=1}^{I} (\theta R_i) n_i = O \; . \tag{17}$$

This same point is made in Section IV C and IV D in a more formal way.

From equation (17) then, sufficient observed numbers for inequality (15) to hold depend entirely upon the person-years in each of the I age groups. One can see that for rare causes of deaths, that is, small specific rates $R_i$ in the standard, more person-years are required to provide the observed numbers that will assure inequality (15) of holding. These can be increased for a fixed initial cohort age distribution by increasing the follow-up time.

Determination of initial cohort size and/or length of follow-up can be empirically arrived at, given a set of standard rated $R_i$, by constructing variations of the anticipated Table 8 and using equation (17) and inequality (15). If in these calculations, the expected numbers given by equation (16) exceed five, then the normal approximation to the Poisson distribution given by the standardization (4) may be used rather than inequality (15).

Note that the sample size/follow-up requirement here flows from the requirement to conclude $O$ is greater than $E$ if in fact $O/E$ is greater than $\theta$. This is somewhat different from, and much simpler than, the usual power calculations, wherein one would specify as the null hypothesis: $H_0$: $E(O) = E$ and the alternative $H_1$: $E(O) \neq E$, with specific alternative of interest $E(O) = \theta E$. A more theoretical approach to sample size determination is presented by Taulbee and Symons (1979), based upon the survivorship ideas sketched in Section IV D. A less detailed discussion is provided by Pell, O'Berg, and Karrh (1978).

IV.  DIFFICULTIES AND ALTERNATIVES

A.  Shortcomings of the SPMR and SMR

Before proceeding with the main difficulty, specifically that of comparing SPMRs and SMRs, the SPMR has a unique problem. When one cause of death, say j', shows an excess,

another cause j is likely to show a deficit.  This is
because the age specific totals, $d_{i+}$, are elevated by extra
cause j' deaths.  Consequently with the calculation (2), the
expected number of deaths for cause j, $e_{ij}$, will be larger
than without the excess of cause j' deaths.  The observed
deaths, $d_{ij}$, will then appear too small.  Put another way,
since the sum (over all causes) of the $P_{ij}$ in (1) is unity,
the observed and expected totals for each age group must be
equal.  A "small" expected number for one cause must be
balanced by an increase in another cause or causes.  Milham
(1975) and Monson (1974) reasonably suggest examining
another cause for significance by restricting the analysis
to a portion of the other causes, such as all cancers, and
excluding an excessive cause that is not of primary
interest, for example accidental deaths.

Returning to the major concern, the SMRs computed for
different industrial groups are generally not comparable.
This point was made by Hill (1959) for standardized
mortality ratios and by Yule (1934) for indirect adjusted
rates.  More recently Miettenen (1972), Chiazze (1976), and
Wong (1977) have reiterated this.

In fact, comparisons are usually meaningless when there
exist differences in the age composition or age-specific
mortality rates, as would typically be expected.  Table 11
presents a simple, contrived illustration where the age-
standardized mortality ratios are unity (or 100%) for two
quite different age specific situations.  A conclusion of

Table 11   Hypothetical Example with Same Age Standardized Mortality Ratios for Two Workforces with Different Age-Specific Experiences

| Age group | Standard rates | Workforce A | | | | Workforce B | | | |
|---|---|---|---|---|---|---|---|---|---|
| | | Person-years | Observed deaths | Expected deaths | Mortality ratio | Person-years | Observed deaths | Expected deaths | Mortality ratio |
| 20–39 | 0.02 | 200 | 3 | 4.0 | 0.750 | 50 | 2 | 1.0 | 2.000 |
| 40–59 | 0.03 | 100 | 3 | 3.0 | 1.000 | 100 | 3 | 3.0 | 1.000 |
| 60+ | 0.04 | 50 | 3 | 2.0 | 1.500 | 200 | 7 | 8.0 | 0.875 |
| Total | | 350 | 9 | 9.0 | 1.000 | 350 | 12 | 12.0 | 1.000 |

1. The mortality ratio in the total row is an age standardized mortality ratio.

similar mortality experience, as suggested by the SMRs, is
clearly erroneous. Difficulties can even come about when

(1)  the workforces have different age distributions
     but have the same age-specific mortality
     rates, or

(2)  the workforces have the same age distributions but
     have different age-specific mortality rates.

Gaffey (1976, p. 158) constructed an example of situation
(2) with SMRs for the two groups being equal.

   Despite these well known limitations, the SMR has been
the summary statistic for occupational mortality cohort
studies. As comparisons will naturally continue to be made,
one recommendation, if followed, would help considerably in
reporting and interpreting SMRs. The only protection
against the pitfalls described in the preceding paragraph is
to always provide, as a part of good reporting, and demand
to see, as a part of careful interpreting, the specific
distributions by the categories used in the standardization,
usually age groups, of

(1)  the workforce size, person-years or number of
         individuals,

(2)  the observed number of deaths, and

(3)  the expected number of deaths.

This is of course not practical for all causes of death
usually examined in an occupational cohort mortality study,
but is reasonable for the one or two specific causes of
special interest and for all causes of death.

Such reporting in the literature would also provide
comparison populations for subsequent use by researchers
conducting mortality studies of other industrial groups.
Pell, O'Berg, and Karrh (1978) provide an excellent example
of this type of reporting.


B.   Direct Adjustment of Mortality Rates

Restricting the choice of alternatives for the SMR to a
single summary statistic, a directly adjusted mortality rate
or comparative mortality index, as described by Hill (1959,
p. 223), avoids some of the limitations of an SMR.  However,
it requires that the workforce is of sufficient size in each
age band (five or 10 years), as discussed in Section III E
6, to have experienced some deaths.  Age-specific death
rates for the study population can then be computed and
these applied to a standard workforce distribution.
Variations in the composition of the workforce are therefore
eliminated, but differences in the pattern of age-specific
death rates may still produce misleading conclusions.

More precisely, consider the following notation.  Let i
index the categories of standardization, usually age groups
for a specified sex-race portion of a workforce.  For
category i then, let $n_i$ be the number of person-years or
number of individuals, $d_i$ be the number of deaths of a
particular cause, and $r_i (=d_i/n_i)$ be the corresponding death
rate, perhaps expressed per 1,000 or 10,000.  For two
populations with rates $r_{1i}$ and $r_{2i}$ in age group i,

respectively, the directly adjusted rate for the first
population is given by

$$\sum_{i=1}^{I} r_{1i}N_i \Big/ \sum_{i=1}^{I} N_i, \tag{18}$$

where $N_i$ is the size of the standard population in age group
i. The same weighted average of the $r_{2i}$ for the second
population gives its directly adjusted rate. With some
simple algebra one can show that

$$\sum_{i=1}^{I} r_{1i}N_i \Big/ \sum_{i=1}^{I} N_i = \sum_{i=1}^{I} r_{2i}N_i \Big/ \sum_{i=1}^{I} N_i \tag{19}$$

implies that $r_{1i} = r_{2i}$, for all I age groups, <u>only</u> when
these age specific rates are in constant ratio or constant
difference for each age group. That is, the equality of
directly adjusted rates (utilizing the same standard) only
implies, or is implied by, the equality of the age-specific
rates when they are in constant ratio or they have constant
difference in each age group.

Some investigators have used this approach to evaluate
the mortality experience of an occupational cohort.
Enterline (1974) used this method to compare cancer
mortality among chromate workers with that among oil
refinery workers as part of a broader study of respiratory
cancer among chromate workers. Mancuso, Ciocco, and El-
Attar (1968) also studied rubber worker mortality in this
way.

There are two major difficulties with applying this
approach to occupational mortality studies. First, as

indicated above, directly adjusted rates also have
limitations as a summary statistic because they reflect only
the adjusted crude mortality and not the age-specific
patterns.  Second, and especially relevant to mortality
studies focusing on a site-specific cancer, several age
groups may have very few or no deaths observed.  The age-
specific rates then are not stable, since they can fluctuate
widely if one death, more or less, is observed in those age
groups.  For rare causes of death the SMR is available, and
therefore used, essentially by default.  However, it should
be remembered that the SMR is based on the total number of
deaths for the cause of interest only and does not use the
distribution of these events over the categories of
standardization.

There are alternative approaches available, however.
Generally these methods require iterative procedures to
estimate the parameters of the statistical models.  The
essence of these approaches will be briefly sketched,
references for more of the details provided, and
improvements expected over the SMR indicated.


C.  Modelling Mortality Rates

As no single mortality index will be able to replace every
schedule of age-specific rates, another approach is to model
the pattern of these rates.  Fortunately some simple models
are known that approximate fairly well the pattern over age
of site-specific cancer rates in man.  Cook, Doll, and

Fellingham (1969) modelled five-year age-specific cancer
rates by the equation

$$r = \alpha t^\beta, \qquad (20)$$

where r is the rate, $\alpha$ and $\beta$ are constants, and t represents
age. The parameters $\alpha$ and $\beta$ can be initially approximated
by fitting a straight line to the I pairs $[\ln(r_i), \ln(t_i)]$,
where $\ln(.)$ denotes natural logarithm and i = 1, 2, ..., I,
$r_i$ is the $i^{th}$ specific rate, and $t_i$ denotes the age, say at
the mid-point, for age group i.

These fitted patterns can then be compared for two or
more industrial populations. As these are straight lines
when ln(rate) is plotted against ln(age), the populations
may differ in slope, intercept, or both. (By dividing age
by 20 or 30, the intercept, exponentiated, is the age-
specific rate at that "zeroing" age.) One can see then with
these two parameters considerable flexibility is available
to handle situations, such as non-proportional age-specific
rates, that pose serious difficulties to any single summary
statistic, for example the SMR or SPMR.

The determination of statistical significance of two
fitted lines is more complex, however. The likelihood
ratio test is recommended and is described in the appendix
of Cook *et al.* (1969) for this application. The estimates
required are maximum likelihood estimates. Presuming that
the number of deaths in age group i, $d_i$, is Poisson
distributed with mean

$$n_i(\alpha t_i^{\;\beta}), \qquad (21)$$

the likelihood is readily formed and standard likelihood methods can be employed to obtain the maximum likelihood estimates by iterative procedures. Methods are available, for example the grid search based algorithm of Kaplan and Elston (1972), to produce these estimates and evaluate the log-likelihoods needed to test for significant differences in the mortality patterns.

Concern for the adequacy of the model for the mortality rates is natural. The representation (20) is known in survivorship analysis as a Weibull hazard. The Gompertz hazard, namely

$$r = \exp (\gamma + \delta t), \tag{22}$$

where $\gamma$ and $\delta$ are parameters, is also used to model mortality rates in human populations, as in Spiegelman (1955, p. 88). Gross and Clark (1975, pp. 14-22) discuss several survivorship distributions and their corresponding hazards. Also described are graphical methods for selecting a hazard function, and thereby a survivorship density.

With a model for the age-specific mortality rates additional interpretation and leverage on previous difficulties is possible for occupational mortality applications. For example, by presuming the model

$$r = \alpha(t - t_o)^{\beta}, \tag{23}$$

Cook $et\ al.$ (1969) estimated a latency, $t_o$, for the age-specific pattern of prostate cancers to be 32.5 years. Doerfler and Symons (1979) offered a quantification of the healthy worker effect using model (20) by comparing

industrial mortality rates with those of a U.S. general
population.

In addition and equally important as having
interpretative flexibility, there are some computational
advantages. One uses the specific rates in all the age
bands to smooth the rates based upon few deaths in the lower
age groups. Also, if a death rate for the cause of interest
at age 20 or 30 say, can be obtained from other studies,
that is, an intercept can be presumed known, then only the
slope parameter needs to be estimated. Site-specific
cancers with a latency period would be especially suited to
modelling in this way.

Restricted versions of the Weibull hazard form (20)
have been related to the standardized mortality ratio and
standardized proportional mortality ratio. Specifically,
Kilpatrick (1962) has identified the SMR as the maximum
likelihood estimate of a constant multiple $\theta$ of each age
specific death rate $R_i$ in the standard by presuming a
Poisson model as in (21), where the number of deaths, $d_i$,
has mean

$$\theta R_i n_i. \qquad (24)$$

This is the same structure presumed with equation (17). In
addition Breslow (1978, 1976 and 1975) has also related the
SMR to similar models. These statistics are shown to be
estimators of well defined single parameter descriptions of
the age-specific pattern of death rates.

The ideas presented here are a generalization of the
SMR, an estimator of only the adjusted crude death rate for

a specific cause or all causes. These models, for example
(20) and (22), allow a summarization of the pattern of age-
specific death rates using the fundamental statistical ideas
of regression. Opportunity to test the adequacy of model
(24), and consequently the SMR, by standard likelihood
methods is straightforwardly available.

The models fitting these patterns are well known in
survivorship analysis as hazard functions. This provides a
direct link to more complete use of the available data by
survivorship methods, which are sketched next.

D.  Survivorship Methods

In producing the age-specific rates, $r_i$, of the previous
section, a partial digestion of the information from the
members of the cohort has taken place. The number of
person-years of experience, $n_i$, via the methods of Section
III B or a tabulation of a mid-year population, has been
determined and the corresponding number of deaths, $d_i$,
tabulated. This is more detailed than relating only the
total number of deaths to the total number expected in an
SMR. However, it does not make as complete use of the
available data as does survivorship analysis.

By way of introduction to these methods, notice that a
schedule of specific death rates can be used to generate the
age-specific probabilities of dying in each age group by
life table methods. The construction of a current life
table involves several steps, as in Spiegelman (1955,

Chapter 5). Fundamental to the process is the conversion of the specific rates to the probabilities of dying during an age-band after surviving to the lower age of the age-band. Various assumptions are employed to make the conversion, one based upon the Gompertz hazard as a model for the specific death rates. It is known as Greville's method, as in Spiegelman (1955, p. 88). Such a life table then provides estimates of life expectancy and probabilities of dying at or surviving to various ages.

Survivorship methods provide the same information as life tables, but based upon explicit use of the information on the individual cohort members using a model rather than the semi-digested information presented as age-specific rates. For each individual we expect to know their age, race, and sex as baseline variables to the follow-up period. From follow-up we know whether or not each individual survived the study period. For those lost to follow-up we know the date of exit. For those dying, we know the date and the cause of death. Deaths from causes not of interest are treated as lost to follow-up.

The model for a survivorship analysis can be determined by specification of the form of the hazard. This is quite flexibly represented by

$$\lambda(t|\underline{z}) = \lambda_0(t)\exp\{\underline{\beta z}\} \tag{25}$$

where $\underline{z}$ is a vector of covariables, for example age, race, and sex designations, at follow-up time $t=0$, the beginning of the study. Feigl and Zelen (1965) first introduced this

form and Cox (1972) popularized it. The vector $\underline{\beta}$ contains
the coefficients to be estimated and $\lambda_0(t)$ is the underlying
hazard at $\underline{z} = 0$. The Weibull or Gompertz hazards of Section
IV C may be utilized for $\lambda_0(t)$. The probability of
surviving until time t for an individual with baseline
covariate vector $\underline{z}$ is given by

$$S(t|\underline{z}) = \exp\{-\int_0^t \lambda(\tau|\underline{z})d\tau\} \qquad (26)$$

and the density, which is the product of the hazard and
survival functions, is given by

$$f(t|\underline{z}) = \lambda(t|\underline{z}) \exp\{-\int_0^t \lambda(\tau|\underline{z})d\tau\}, \qquad (27)$$

as in Gross and Clark (1975).

The likelihood of data from a cohort mortality study
can then be written as follows:

$$L[\underline{t}|\underline{z}_i\text{'s}, \underline{\beta}, \lambda_0(t_i)\text{'s}] = \prod_{\substack{\text{all} \\ \text{deaths}}} \lambda(t_i|\underline{z}_i) \prod_{\substack{\text{all} \\ \text{individuals}}} S(t_i|\underline{z}_i), \qquad (28)$$

where $\underline{t}$ is the vector of times to death for those dying, or
time followed without death due to the cause of interest.
These latter times are the time lost to follow-up, time to
death from another cause, or time of observation if censored
by the end of the study. Iterative calculations are usually
required to obtain the maximum likelihood estimates. These
estimates are used in testing hypotheses by the likelihood
ratio criterion.

A more comprehensive description of survival analysis
methods is given in Chapter 4. An example of a cohort
mortality study analysed by these methods is described by
Taulbee (1977, pp. 54-61). The analysis allows a solid

description of the mortality experience of active white male rubber workers, reported earlier by Andjelkovic *et al.* (1976), but of course the analysis is still limited by the lack of a comparison population other than the U.S. general population. A test of the null hypothesis of no effect for the number of years worked, after adjusting for age, was not rejected for this cohort, an example of internal comparisons possible with these methods. There are obvious difficulties with variables, such as years worked, which is strictly less than age even after subtracting 15 or so years, and is also highly correlated with age. Only with better data on employed populations, analyzed by methods such as these, will these inherent difficulties be overcome.

From the survivorship methods statements of relative risk sometimes can be made. Relative risk is the ratio of the probability of dying of a specific cause given a particular exposure to the same probability without the exposure. This is an abbreviated expression of a dose-response relationship unless the relative risk is given as a function of the levels of exposure, for example by a logistic regression, as by Cox (1972). However desirable as an end-product of an industrial cohort mortality study, documenting a dose-response relationship is seldom possible as information on the level of exposures is usually not available.

We note however that there is a tendency to relate SMRs and relative risk. See for example Wong's (1977) discussion

of Enterline (1976). When the age-specific death rates are
small and in constant ratio across the (short) age bands,
the SMR numerically approximates relative risk. However,
there are two distinct concepts involved and we abstract a
more thorough discussion presented by Symons and Taulbee
(1981). The SMR is an index of the death rate and relative
risk is a measure of the probability of dying. They are
functionally related through equation (26). If we let $r_{1i}$
be the death rate for age group i in the exposed population
and similarly $r_{2i}$ in the unexposed population, then the
mortality ratio is $r_{1i}/r_{2i}$. Let $q_{1i}$ be the conditional
probability of death for an individual of age i in time t
for the exposed group and similarly $q_{2i}$ in the unexposed
group. Then $q_{1i}$ is equal to $1 - \exp\{-r_{1i}t\}$, from (26),
assuming a constant hazard for age group i. When the death
rate $r_{1i}$ is small and time t is short, $q_{1i}$ is approximately
equal to $r_{1i}t$, by the Taylor series expansion of the
exponential function. By definition the relative risk of
death at age i in time t with the exposure is $q_{1i}/q_{2i}$, or
approximately the mortality ratio $r_{1i}/r_{2i}$ when the death
rates are small and time t is short. If the age-specific
death rates are in constant proportion for all ages, or
equivalently the proportional hazards assumption of
survivorship analysis holds with age as a covariate, the
relative risk can be reasonably approximated by the SMR.
This point does not seem well recognized.

    To conclude this section, the CMH procedure described
earlier in connection with proportional mortality also has

application to the comparison of survival experience in two

occupational cohorts. We emphasize that the data for each

population needs to be in "follow-up form", that is, as

described in Section III B. Wong (1977) gives the layout of

the data for this analysis. The application is sketched by

Mantel (1963) and elaborated upon in Mantel (1966).

The CMH procedure is extremely powerful when the

survival in the cohorts is consistently different at all

follow-up time intervals and for all levels of concomitant

variables. Peto and Peto (1972) document this fact.

However, when the proportional hazards assumption, indicated

by the constant multiple in equations (17) and (24), is not

operative, other methods are required. Taulbee (1979)

provides a test of this assumption and an alternative

approach in the event that the proportional hazards

assumption is not satisfied.


V.   SUMMARY AND DISCUSSION


A.   Plight of Single Summary Statistics

Any single dimensional summary statistic for cohort

mortality studies, like the standardized mortality ratio or

standardized proportional mortality ratio, will fail in

attempts to represent diverse patterns of age-specific

mortality rates. Breslow (1978) makes this point quite

clearly by providing multiplicative models involving a

single parameter that suggest the SMR or SPMR as an initial

estimate for a well defined parameter in models such as
those represented by equation (21).

The standardized mortality ratio is limited in its
ability to represent patterns of age-specific mortality
rates, but for many situations it probably will not be
misleading. Examples where it fails are largely contrived;
see for example our Table 11 and panel B of Table 1 in
Gaffey (1976). Both involve reversals of the specific
mortality rates, situations known to be poorly described by
a synoptic mortality index. Difficulties also can occur in
practice as Kilpatrick (1963) shows, but these instances are
apparently rare. The use of the SMR, together with an
examination of the mortality ratio calculations within age
groups should be a safe, informative practice.

Modelling the schedule of death rates, as described in
Section IV C, or survivorship analysis, as indicated in
Section IV D, are alternatives that require more complex
statistical approaches but offer flexibility in handling
some of the methodological problems. Nevertheless,
statistical sophistication can never compensate completely
for the lack of better comparative data.


B. Concern for Overall Level of Significance

Several causes of death are often examined in an occupational cohort study. The overall level of significance of the study then shoud be of concern. By virtue
of examing several hypotheses for stat-stical

significance, some basic precautions to protect the overall
level of significance are warranted. The simplest of these
is to use the Bonferroni inequality, which guarantees an
overall level of significance of no more than $\alpha$ if each of
the k tests to be performed is carried out at the $\alpha/k$ level.

Without an attempt by a researcher to control the
overall level of significance, a reader is best advised to
consider a reported statistical significance as the possible
result of considering a large number of tests, each
performed at the $\alpha$ level. Consistency of such a result with
those reported by others, that is, some form of replication
of the claimed significant result, should be required to put
it in proper perspective. Mantel and Haenszel (1959)
present this position very well.

REFERENCES

Andjelkovic, D., Taulbee, J.D., and Symons, M.J. (1976).
       Mortality experience of a cohort of rubber workers,
       1964-1973. *Journal of Occupational Medicine*,
       18:387-94.

Andjelkovic, D., Taulbee, J.D., Symons, M.J., and Williams,
       T. (1977). Mortality of rubber workers with
       reference to work experience. *Journal of
       Occupational Medicine*, 19:397-405.

Breslow, N.E. (1975). Analysis of survival data under the
       proportional hazards model. *International
       Statistical Review*, 43:45-58.

Breslow, N.E. (1976). Some statistical models useful in the
       study of occupational mortality. Proceedings of a
       Conference on Environmental Health, Alta, Utah.
       July 5-9. p. 88-103.

Breslow, N.E. (1978). The proportional hazards model:
       applications in epidemiology. *Communications in
       Statistics: Theory and Methods*, A7(4):315-332.

Chiazze, L. (1976). Problems of study design and
    interpretation of industrial mortality experience.
    *Journal of Occupational Medicine*, 18:169-170.

Cochran, W.F. (1954). Some methods for strengthening the
    common chi-square tests. *Biometrics*, 10:417-450.

Cook, P.J., Doll, R. and Fellingham, S.A. (1969). A
    mathematical model for the age distribution of
    cancer in man. *International Journal of Cancer*,
    4:93-112.

Cornfield, J. (1951). A method of estimating comparative
    rates from clinical data. Applications to cancer of
    the lung, breast, and cervix. *Journal of the
    National Cancer Institute*, 14:1269-1275.

Cox, D.R. (1972). Regression models and life-tables (with
    discussion). *Journal of the Royal Statistical
    Society, Series B*, 34:187-202.

Decoufle, P., Thomas, T.L., and Pickle, L.W. (1980).
    Comparison of the proportionate mortality ratio and
    standard mortality ratio risk measures. *American
    Journal of Epidemiology*, 111:263-269.

Doerfler, D.L. and Symons, M.J. (1979). Restricted
    multiplicative model for occupational mortality
    rates. Presented at The Biometric Society Spring
    Meetings. New Orleans, Louisiana. April.

Enterline, P.E. (1974). Respiratory cancer among chromate
    workers. *Journal of Occupation Medicine*,
    16:523-526.

Enterline, P.E. (1976). Pitfalls in epidemiological
    research, an examination of the asbestos literature.
    *Journal of Occupational Medicine*, 18:150-156.

Eighth Revision International Classification of Diseases,
    Adapted for Use in the United States (1968). *U.S.
    Department of Health, Education, and Welfare.*
    Public Health Service. National Center for Health
    Statistics. Publication No. 1693.

Feigl, P. and Zelen, M. (1965). Estimation of exponential
    survival probabilities with concomitant information.
    *Biometrics*, 21:826-838.

Gaffey, W.R. (1976). A critique of the standardized
    mortality ratio. *Journal of Occupational Medicine*,
    18:157-160.

Grizzle, J.E., Starmer, C.F., and Koch, G.G. (1969).
    Analysis of categorical data by linear models.
    *Biometrics*, 25:489-504.

Gross, A.J. and Clark, V.A. (1975). *Survival Distributions:
    Reliability Applications in the Biomedical Sciences.*
    John Wiley and Sons, New York.

Hill, A.B. (1959). *Principles of Medical Statistics.*
    Oxford University Press, New York. Sixth Edition.

Hill, I.D. (1972). Computing man years at risk. *British
    Journal of Preventive and Social Medicine,*
    26:132-134.

Johnson, N.L. and Kotz, S. (1969). *Distributions in
    Statistics: Discrete Distributions.* John Wiley and
    Sons, Inc., New York.

Kaplan, E.B. and Elston, R.C. (1972). A package for maximum
    likelihood estimation (MAXLIK). *Institute of
    Statistics Mimeo Series No.* 823. Department of
    Biostatistics, School of Public Health, University
    of North Carolina, Chapel Hill, N.C.

Kilpatrick, S.J. (1962). Occupational mortality indices.
    *Population Studies,* 16:175-182.

Kilpatrick, S.J. (1963). Mortality comparisons in socio-
    economic groups. *Applied Statistics,* 12:65-86.

Kupper, L.L., McMichael, A.J., Symons, M.J., and Most,
    B.M. (1978). On the utility of proportional
    mortality analysis. *Journal of Chronic Diseases,*
    31:15-22.

Li, F.P., Fraumeni, J.F., Jr., Mantel, N., and Miller,
    R.W. (1969). Cancer mortality among chemists.
    *Journal of the National Cancer Institute,*
    43:1159-1164.

Lloyd, J.W. and Ciocco, A. (1969). Long-term mortality
    study of steel workers. *Journal of Occupational
    Medicine,* 11:299-310.

Lloyd, J.W., Lundin, F.E., Jr., Redmond, C.K., and
    others. (1970). Long-term mortality study of steel
    workers: IV. mortality by work area. *Journal of
    Occupational Medicine,* 12:151-157.

Mancuso, T.F., Ciocco, A., and El-Attar, A.A. (1968). An
    epidemiological approach to the rubber industry:  a
    study based on departmental experience. *Journal of
    Occupational Medicine,* 10:213-232.

Mantel, N. (1963). Chi-square tests with one degree of freedom: extensions of the Mantel-Haenszel procedure. *Journal of the American Statistical Association*, 58:690-700.

Mantel, N. (1966). Evaluation of survival data and two new rank order statistics arising in its consideration. *Cancer Chemotherapy Reports*, 50:163-170.

Mantel, N. and Haenszel, W. (1959). Statistical aspects of the analysis of data for retrospective studies of disease. *Journal of the National Cancer Institute*, 22:719-748.

Marsh, G.M. and Enterline, P.E. (1979). A method for verifying the completeness of cohorts used in occupational mortality studies. *Journal of Occupational Medicine*, 21:665-670.

Marsh, G.M. (1980). OCMAP: a user-oriented occupational cohort mortality analysis program. *The American Statistician*, 34:245.

McMichael, A.J. (1976). Standardized mortality ratios and the healthy worker effect: Scratching beneath the surface. *Journal of Occupational Medicine*, 18:165-168.

Miettinen, O.S. (1972). Standardization of risk ratios. *American Journal of Epidemiology*, 96:383-388.

Milham, S., Jr. (1975). Methods in occupational mortality studies. *Journal of Occupational Medicine*, 17:581-585.

Milham, S., Jr. (1978). Experience in using death certificate occupational information. Presented at the Public Health Conference on Records and Statistics. Washington, D.C., June.

Monson, R.R. (1974). Analysis of relative survival and proportional mortality. *Computers and Biomedical Research*, 1:325-332.

Monson, R.R. and Nakano, K.K. (1976). Mortality among rubber workers. I. White union employees in Akron, Ohio. *American Journal of Epidemiology*, 103:284-296.

Pagnoto, L.D., Elkins, H.B., and Brugsch, H.G. (1979). Benzene exposure in the rubber-coating industry - a follow-up. *American Industrial Hygiene Association Journal*, 40:137-146.

Pell, S., P'Berg, M.T., and Karrh, B.W. (1978). Cancer
    epidemiology surveillance in the Du Pont Company.
    *Journal of Occupational Medicine*, 20:725-740.

Pell, S. (1978). Mortality of workers exposed to
    chloroprene. *Journal of Occupational Medicine*,
    20:21-29.

Peto, R. and Peto, J. (1972). Asymptotically efficient rank
    invariant test procedures (with discussion).
    *Journal of the Royal Statistical Society* A
    135:185-207.

Peto, R. and Pike, M.C. (1973). Conservatism of the
    approximation in the logrank test for survival data
    or tumor incidence data. *Biometrics*, 29:579-583.

Redmond, C.K. and Breslin, P.P. (1975). Comparison of
    methods for assessing occupational hazards. *Journal
    of Occupational Medicine*, 17:313-317.

Shyrock, H.S., Siegel, J.S., and Associates. (1973). The
    method and materials of demography. *U.S. Department
    of Commerce, Bureau of the Census*. Volume
    2. Chapter 15.

Spiegelman, M. (1955). *Introduction to Demography*. The
    Society of Actuaries, Chicago. Chapter 5.

Symons, M.J. and Taulbee, J.D. (1981). Practical
    considerations for approximating relative risk by
    the standardized mortality ratio. *Journal of
    Occupational Medicine*, 23:413-416.

Taulbee, J.D. (1977). A general model for the hazard rate
    with covariables and methods for sample size
    determination for cohort studies. *Institute of
    Statistics Mimeo Series*. No. 1154. Department of
    Biostatistics, University of North Carolina at
    Chapel Hill.

Taulbee, J.D. (1979). A general model for the hazard rate
    with covariables. *Biometrics*, 35:439-450.

Taulbee, J.D. and Symons, M.J. (1979). Approximate sample
    size requirements for cohorts studies with
    covariables. Presented at the Joint Meeting of the
    American Statistical Association, Biometric Society
    and Institute for Mathematical Statistics,
    Washington, D.C. August. Revision to appear in
    *Biometrics*.

Wong, O. (1977). Further criticisms on epidemiological
    methodology in occupational studies. *Journal of
    Occupational Medicine*, 19:220-222.

Yule, G.U. (1934). On some points relating to vital
    statistics, more especially statistics of
    occupational mortality (with discussion). *Journal
    of the Royal Statistical Society*, 94:1-84.

# 3
# STATISTICS OF CASE-CONTROL STUDIES

Norman E. Breslow
University of Washington
Seattle, Washington

Nicholas E. Day
International Agency for Research on Cancer
Lyon, France

## I. INTRODUCTION

The case-control or retrospective study is one of the most
important tools currently available to the chronic disease
epidemiologist for investigation of cancer etiology. New
cases of a particular site and/or type of cancer occurring
in a defined population are compared to a control sample

Much of the material in this chapter is drawn from a
monograph published by the International Agency for Research
on Cancer (IARC). For a fuller discussion of the concepts
involved, detailed formulae and worked examples, the reader
should consult Breslow and Day (1980).

Research supported in part by grant 1 K07 CA00723
from the United States National Cancer Institute. The
assistance of the Alexander von Humboldt foundation in
preparing the revision is gratefully acknowledged.

with respect to genetic markers, exposure to environmental
agents, and other suspected risk factors. The advantages of
this study design are that a relatively large number of
cases may be ascertained in a short period of time and that
information may be collected simultaneously on a variety of
nuisance factors which may confound the association of
primary interest. Disadvantages relate to the accuracy of
the exposure histories, which must usually be obtained by
interview or other retrospective means, and to questions of
bias in the selection of the control sample.

As Mantel and Haenszel (1959) noted in their landmark
paper on the case-control study, "a primary goal is to reach
the same conclusions ... as would have been obtained from a
forward study, if one had been done." This viewpoint is
fundamental to the methodological approach suggested here.
We first consider statistical methods appropriate for the
forward or prospective study, in which the probability of
disease development is related to each individual's exposure
history, and then show that precisely the same methods of
analysis may be used also with the case-control design.

## II.  MEASURES OF DISEASE/RISK FACTOR ASSOCIATION:   THE
   RELATIVE RISK

Quantitative studies of risk factors and disease require a
measure of exposure to the risk variables, as well as one of
disease occurrence, and some method of associating the two.
Risk variables may be difficult to measure precisely, since

individual histories differ widely with respect to the onset, duration and intensity of exposure, and whether it was continuous or intermittent. Nevertheless it is often possible to make a crude classification into an exposed versus a non-exposed group, for example by comparing confirmed cigarette smokers with lifelong non-smokers or urban with rural residents. In cancer epidemiology the most appropriate measure of disease is incidence, informally defined here as the probability that a disease-free person becomes ill during a specified time interval. The incidence of particular cancers varies remarkably according to a wide range of factors including age, sex, calendar time, geography and ethnicity. Thus, whatever measure is used to associate incidence with exposure, this may also vary with age, sex and time. It is desirable to have a measure of association which is as stable as possible since there is then greater justification for expressing the effect of exposure in a single summary number. The more the measure varies, the greater is the need to describe how the effect of exposure interacts with, i.e., is modified by, demographic and other extraneous factors.

Relative risk, defined as the ratio of incidences in the exposed vs. the non-exposed group, has several advantages as a measure of association. First, empirical studies of cancer have shown that it often comes reasonably close to meeting the ideal of stability over a wide range of ages, times and places. Second, its use allows some

judgement to be made of the extent to which the observed
association may be due to confounding factors not accounted
for in the analysis.  As Cornfield, Haenszel, Hammond,
Lilienfeld, Shimkin and Wynder (1959) noted, in order that
confounding explain a relative risk of ten, the confounding
factor (i) must be at least ten times more prevalent in the
exposed group and (ii) must itself increase the risk by at
least tenfold.  Finally, and what is especially important in
the present context, the relative risk is in principle
directly estimable from case-control studies, as we shall
demonstrate shortly.

    None of these three properties is enjoyed by the excess
risk, defined as the difference in incidences between
exposed and non-exposed.  Nevertheless the excess risk often
gives a better impression of the public health importance of
the association (Berkson, 1958).  Thus, even though the
statistical analysis may be oriented towards estimation of
the relative risk, it is important to combine the results
obtained with whatever information may be available about
actual incidence in order to better interpret their
significance.


III.   PROSPECTIVE VS. RETROSPECTIVE SAMPLING IN A 2x2 TABLE

The most satisfactory case-control studies are those in
which cases are identified through a cancer registry or some
other system which covers a well-defined population.  For
hospital-based studies the referent population, consisting

of all those "served" by the given hospital, may be more
imaginary than real. The sample will commonly consist of
all new cases arising during the study period, or at least
all which were successfully interviewed. Otherwise it is
assumed to be a random sample of the actual cases.
Similarly, the controls are assumed to represent a random
sample of those persons who remain disease-free, though
otherwise at risk, throughout the study period. The control
sample may be a stratified one, so that it has roughly the
same age and sex distribution as the cases. Or controls may
be individually matched to cases on the basis of family
membership, place of residence or other characteristics.
Then they are assumed to constitute a random sample from
within each of the sub-populations formed by the
stratification or matching factors.

   If unlimited resources were available, one would
ideally conduct a prospective investigation of the entire
population. Persons would be classified at the beginning of
the study period on the basis of exposure to the risk
factor, and at the end according to whether or not they had
developed the disease. Suppose that a proportion p of the
individuals at risk in a particular stratum were exposed at
the beginning of the study. Denote by $P_1$ the probability
that an exposed person in this stratum develops the disease
during the study period, and by $P_0$ the analogous quantity
for the unexposed. Then the proportions of individuals who
fall into each of the resulting four categories or cells
are:

|              | Exposed | Non-exposed | Total |
|--------------|---------|-------------|-------|
| Diseased     | $pP_1$  | $qP_0$      | $pP_1+qP_0$ |
| Disease-free | $pQ_1$  | $qQ_0$      | $pQ_1+qQ_0$ |
| Total        | $p$     | $q$         | $1$   |

, (1)

where $q = 1-p$ and $Q_i = 1-P_i$.

For reasonably short studies, meaning on the order of a year of two for most cancers and other chronic diseases, the probabilities $P_1$ and $P_0$ will be quite small, so that $Q_1 \simeq Q_0 \simeq 1$. It follows that the relative risk (RR) is well approximated by the odds ratio $\Psi$ of disease probabilities, i.e.

$$RR = \frac{P_1}{P_0} \simeq \frac{P_1 Q_0}{Q_1 P_0} = \Psi \quad . \tag{2}$$

As Cornfield (1951) observed, this approximation provides the key link between forward and case-control studies *vis-à-vis* estimation of the relative risk. If a sample drawn at random from the entire population were kept under observation for the study period, one could make separate estimates of the three quantities $p$, $P_1$ and $P_0$. Sampling of a cohort of subjects on the basis of their exposure history at the start of the study period, and subsequent surveillance, would allow estimation of $P_1$ and $P_0$. However, in view of the relative rarity of the disease such prospective designs necessitate keeping track of large numbers of individuals. With the retrospective approach, cases of disease are sampled as they arise in the population

at risk and their exposures are compared with a group of
controls drawn at random from those who remain disease-free.
From the samples of cases and controls one may estimate the
following exposure probabilities given disease status:

$$p_1 = pr(Exposed|Diseased) = \frac{pP_1}{pP_1 + qP_0}$$

and                                                              (3)

$$p_0 = pr(Exposed|Disease\text{-}free) = \frac{pQ_1}{pQ_1 + qQ_0}.$$

An easy calculation shows that the odds ratio calculated
from the exposure probabilities is identical to the odds
ratio of the disease probabilities, i.e.

$$\psi = \frac{p_1 q_0}{p_0 q_1} = \frac{P_1 Q_0}{P_0 Q_1}$$          (4)

where $q_i = 1-p_i$ for $i = 0$ and 1. Consequently the relative
risk, as approximated by the disease odds ratio, is directly
estimable from a case-control study even though the latter
provides no information about the absolute magnitude of risk
in the exposed and non-exposed subgroups.

The actual data collected in an epidemiological study,
whether of the prospective or retrospective variety, may be
summarized as frequencies in a series of 2x2 tables, one for
each stratum

|              | Exposed | Non-exposed |       |
|--------------|---------|-------------|-------|
| Diseased     | a       | b           | $n_1$ |
| Disease-free | c       | d           | $n_0$ |
|              | $m_1$   | $m_0$       | N     |

.(5)

For the forward study the marginal totals $m_1$ and $m_0$ are regarded as fixed, so that the likelihood of the data is the product of two binomial distributions with parameters $(P_1, m_1)$ and $(P_0, m_0)$. With the case-control design the marginals $n_1$ and $n_0$ are fixed. The likelihood is again the product of two binomials, but this time the parameters $(p_1, n_1)$ and $(p_0, n_0)$ involve the exposure probabilities.

In order to make inferences about the key parameter $\Psi$ which are not complicated by the other (nuisance) parameters in the model, it is appropriate to consider the conditional distribution of the data assuming all the marginal totals are fixed (Cox and Hinkley, 1974). This is the non-central hypergeometric distribution

$$\mathrm{pr}(a \mid n_1, n_0, m_1, m_0; \Psi) \;=\; \frac{\binom{n_1}{a}\binom{n_0}{m_1-a} \Psi^a}{\sum\limits_{u} \binom{n_1}{u}\binom{n_0}{m_1-u} \Psi^u} \tag{6}$$

where the summation is understood to range over all values $u$ for the number of exposed cases which are consistent with the observed marginals, namely $0, n_1-m_0 \leq u \leq m_1, n_1$. Good summaries of "exact" methods of inference based on (6) are given by Gart (1971; 1979). The fact that the likelihood remains the same upon interchanging the roles of $n$ and $m$ confirms that it arises from either forward or case-control sampling schemes. Consequently, whether they were sampled according to a prospective or retrospective design, the data (5) are analyzed in precisely the same fashion. In the sequel we will observe that the same principle applies also to more complicated designs involving matched sampling and the evaluation of multiple risk factors.


IV.  STRATIFIED LOGISTIC REGRESSION

While perhaps adequate in a few special situations, the dichotomization of a risk factor into the two levels exposed *vs.* unexposed generally represents an oversimplification and tends to obscure important information, for example about dose-response relationships.  Moreover, most epidemiological investigations collect data on several risk factors whose joint effects on disease should be analyzed.  It it also necessary to adjust the relative risk estimates for the presence of confounding or nuisance variables, such as sex, age and ethnicity, which are associated with the exposures and whose causal influence on disease is conceded *a priori*.  Consequently, a generalization of the simple 2x2 table analysis is needed to estimate multivariate relative risk functions in the presence of confounders.

An approach which satisfies these requirements is to model the exposure main effects and interactions in the linear logistic equation (Cox, 1970), while adjusting for the effects of the nuisance factors by stratification.  This allows considerable flexibility in the choice of factors which are controlled via stratification *vs.* those whose effects are modelled.  Suppose that the population has been

divided into I strata on the basis of the nuisance factors
and that a vector $\underline{x}$ of exposures has been recorded for each
individual.  Denote by y a binary variable which takes the
values y = 1 for persons who develop the disease and y = 0
for those who remain disease-free.  Then the model states

$$pr(y=1|\underline{x}, \text{ stratum i}) = \frac{\exp(\alpha_i + \underline{\beta}'\underline{x})}{1+\exp(\alpha_i + \underline{\beta}'\underline{x})}, \qquad (7)$$

where the $\alpha_i$ are nuisance parameters associated with the
strata effects while $\underline{\beta}$ are regression coefficients for the
exposures of interest.

If there is but a single stratum, and a single binary
risk variable coded x = 1 for exposed and x = 0 for
unexposed, then (7) is a special case of the disease
probability model (1) with $\alpha = \log(P_0/Q_0)$ and $\beta = \log(\Psi)$.
More generally, the relative risk in stratum i for persons
with two sets of exposures $\underline{x}*$ and $\underline{x}$ is approximated by the
ratio of odds calculated from (7), namely

$$RR = \exp\{\underline{\beta}'(\underline{x}*-\underline{x})\}. \qquad (8)$$

A large number of possible relationships may be represented
in this form by including among the x's both discrete and
continuous measurements, transformations of such
measurements, and cross-product or interaction terms.
Interactions involving the product of exposures and
stratification variables, for example, allow one to
concisely model and test for possible variations in relative
risk across strata.

If the study covers more than a year or two, the first factor considered for strata formation might be calendar time. Further subdivision may be based on the age of the subject during each time interval and other nuisance factors as appropriate. A notable feature of such stratification is that the subject may progress from one stratum to another as he is followed forward. Also, the probabilities in (7) become conditional probabilities of developing disease in a particular time interval for persons disease-free at its start. A continuous time analog is obtained by allowing the time intervals (strata) to become infinitesimally small. This results in Cox's (1972) proportional hazards model, whereby the ratio of instantaneous age and time specific incidence rates for persons with covariates $\underline{x}*$ and $\underline{x}$ is given exactly by the right hand side of (8). The proportional hazards approach thus has the conceptual advantage of eliminating the odds ratio approximation (2) and obviating the rare disease assumption. Strictly speaking it also requires that each case be compared only with persons who are disease-free at the precise moment his disease is diagnosed. This implies that the controls for each case are chosen to be disease-free at that moment, a requirement which in fact accords well with the actual conduct of most matched case-control studies. See Prentice and Breslow (1978) for a more detailed account of the proportional hazards model as it relates to retrospective designs.

V.   ADAPTATION OF THE LOGISTIC MODEL FOR CASE-CONTROL
     DESIGNS

According to the logistic model just defined, the exposures
x are regarded as fixed quantities while the response
variable y is random.  This fits the forward study precisely
because one does not know in advance whether or not, or
when, a given individual will develop the disease.  With the
case-control design, on the other hand, it is the history of
exposures which should properly be regarded as the random
outcome.  Thus it is important to determine how the logistic
model for disease probabilities, which has such a simple and
desirable interpretation *vis-à-vis* the relative risk, can be
adapted for use with case-control sampling.  We have already
seen for the case of a single binary risk factor that
precisely the same calculations as applied to forward data
are also used with case-control studies to estimate the
relative risk.  In fact this identity of inferential
procedures is a fundamental property of the logistic model.

     We illustrate this feature of the logistic model with a
simple calculation due to Mantel (1973) (see also Seigel and
Greenhouse, 1973), which lends a good deal of plausibility
to the argument.  Suppressing the stratum indicator i for
the time being, denote by $\pi_1$ and $\pi_0$ the sampling fractions
for cases and controls, respectively.  Thus if the indicator
variable z denotes whether (z = 1) or not (z = 0) someone is
sampled, we have $\pi_1 = pr(z=1|y=1)$ and $\pi_0 = pr(z=1|y=0)$.
Typically $\pi_1$ will be near unity, i.e. most potential cases

are sampled for the study, while $\pi_0$ is of a lower order of magnitude.

Consider now the probability that a person is diseased, given that he has exposures $\underline{x}$ and that he was sampled for the case-control study. Using Bayes' theorem, we compute

$$pr(y=1|z=1,\underline{x}) = \frac{pr(z=1|y=1,\underline{x})pr(y=1|\underline{x})}{pr(z=1|y=0,\underline{x})pr(y=0|\underline{x})+pr(z=1|y=1,\underline{x})pr(y=1|\underline{x})}$$

$$= \frac{\pi_1 \exp(\alpha + \underline{\beta}'\underline{x})}{\pi_0 + \pi_1 \exp(\alpha + \underline{\beta}'\underline{x})}$$

$$= \frac{\exp(\alpha* + \underline{\beta}'\underline{x})}{1 + \exp(\alpha* + \underline{\beta}'\underline{x})} ,$$

where $\alpha* = \alpha + \log(\pi_1/\pi_0)$. Consequently the disease probabilities for persons in the sample continue to be given by the logistic model with precisely the same value for the relative risk parameters $\underline{\beta}$ albeit a different constant term. This observation alone would suffice to justify the application of (7) to case-control data provided we could also assume the probabilities for different individuals were independent. However, unless a separate decision was made for each case or control on whether or not to include him in the sample, this will not be true. Some slight dependencies are introduced by the fact that the total numbers of cases and controls are fixed by design, and recourse is therefore needed to a somewhat deeper theory.

One assumption made implicitly in the course of the above derivation deserves emphasis, namely that the sampling probabilities $\pi$ within each stratum depend only on disease

status and not on the exposures. Otherwise there is no
simple relationship between the prospective and
retrospective probabilities, and biassed estimates of
relative risk will result. This is a particular problem
with hospital based studies, since the controls are selected
to have diseases other than cancer which may themselves be
related to the exposures of interest.

Since case-control studies typically involve separate
samples of fixed size from the diseased and disease-free
populations, the independent probabilities are those of the
exposures $\underline{x}$ given disease status. With $n_1$ cases and $n_0$
controls the likelihood is a product of $n_1$ terms of the form
$pr(\underline{x}|y=1)$ and $n_0$ of $pr(\underline{x}|y=0)$. Each of these may be
expressed

$$pr(\underline{x}|y) = \frac{pr(y|\underline{x})pr(\underline{x})}{pr(y)} \qquad (9)$$

as the product of the conditional probabilities of disease
given exposure, specified by the logistic model, times the
ratio of unconditional probabilities for exposure and
disease. How one approaches the estimation of the relative
risk parameters $\underline{\beta}$ from this likelihood depends on whether
the $\underline{x}$ variables themselves, without knowledge of the
associated $y$'s, contain any information about the relative
risk. Such information would be expressed mathematically
through dependence of the marginal distribution $pr(\underline{x})$ on $\underline{\beta}$,
as well as on other parameters, in which case better
estimates of $\underline{\beta}$ could in principle be obtained using the

entire likelihood (9) rather than from the portion of the likelihood specified by (7).

An example where the $\underline{x}$'s were assumed to contain information on their own comes from the earliest applications of the logistic model to studies of coronary heart disease (Truett, Cornfield and Kannel, 1967). Supposing the $\underline{x}$'s to have multivariate normal distributions within each disease category with means $\underline{u}_1$ for cases, $\underline{u}_0$ for controls and a common covariance matrix $\sharp$, these authors noted that the relative risk parameters could be written

$$\underline{\beta} = \sharp^{-1}(\underline{u}_1 - \underline{u}_0) . \qquad (10)$$

Estimation of $\underline{\beta}$ from the full likelihood is then a by-product of normal theory discriminant analysis. However if the assumption of multivariate normality does not hold, severe bias can result from this approach, and it is consequently not recommended (Halperin, Blackwelder and Verter, 1971; Efron, 1975; Press and Wilson, 1978).

A more prudent course is to allow the marginal distribution $pr(\underline{x})$ to remain completely arbitrary. Quite remarkably, it follows that the joint estimation by maximum likelihood of $pr(\underline{x})$ and $\underline{\beta}$ from (9) yields estimates and asymptotic standard errors for $\underline{\beta}$ which are identical to those based on the disease probability model (7) (Anderson, 1972; Prentice and Pyke, 1979). This result justifies fully the basic principle that one applies the same inferential techniques to case-control data as would be applied to forward data from the same population.

VI.  CONDITIONAL LOGISTIC REGRESSION

An alternative but related approach to making inference from
(9) is to eliminate the nuisance term $\text{pr}(\underline{x})/\text{pr}(y)$ entirely
by consideration of an appropriate conditional likelihood,
just as was done earlier for the 2x2 table.  Suppose that a
particular stratum of $n = n_1 + n_0$ subjects yields the exposure
vectors $\underline{x}_1, \ldots, \underline{x}_n$, but that it is not specified which
pertain to the cases and which to the controls.  The
conditional probability that the first $n_1$ $\underline{x}$'s in fact go
with the cases, as observed, and the remainder with the
controls, may be written

$$\frac{\displaystyle\prod_{j=1}^{n_1} \text{pr}(\underline{x}_j|y=1) \prod_{j=n_1+1}^{N} \text{pr}(\underline{x}_j|y=0)}{\displaystyle\sum_{\ell} \prod_{j=1}^{n_1} \text{pr}(\underline{x}_{\ell_j}|y=1) \prod_{j=n_1+1}^{N} \text{pr}(\underline{x}_{\ell_j}|y=0)} \quad ,$$

where the denominator sum is over all $\binom{N}{n_1}$ possible ways of
dividing the numbers from 1 to n into one group $\{\ell_1, \ldots, \ell_{n_1}\}$
of size $n_1$ and its complement $\{\ell_{n_1+1}, \ldots, \ell_N\}$.  Substituting
from (7) and (9) this reduces to

$$\frac{\displaystyle\prod_{j=1}^{n_1} \exp(\underline{\beta}'\underline{x}_j)}{\displaystyle\sum_{\ell} \prod_{j=1}^{n_1} \exp(\underline{\beta}'\underline{x}_{\ell_j})} \quad , \tag{11}$$

which depends only on the parameters of interest (Liddel,
McDonald and Thomas, 1977; Prentice and Breslow, 1978).

The full likelihood consists of a product of terms (11), one for each stratum. It has precisely the same form as Cox's (1975) partial likelihood, used to analyze data from a forward study under the proportional hazards model. However the "risk sets" in the denominator, instead of containing all persons in the study who remain disease-free at the time of diagnosis of the corresponding case, instead consist merely of the cases and matched controls actually sampled at that moment.

Computer programs for making likelihood inferences about $\underline{\beta}$ on the basis of (11) have been written and are available (Smith, Pike, Hill, Breslow and Day, 1981; Gail, Lubin and Rubinstein, 1981). Through use of recursive formulae, the latter program permits strata containing moderate numbers of both cases and controls to be efficiently analyzed. However the full conditional analysis is usually restricted for use with matched controls, or for similar situations where the stratification is very fine. For example, when the $i^{th}$ stratum consists of but a single case with exposures $\underline{x}_{i0}$, and $M_i$ controls with exposures $\underline{x}_{i1}, \ldots, \underline{x}_{i,M_i}$, the likelihood takes the simple and readily computed form

$$\prod_{i=1}^{I} \frac{1}{1+\sum\limits_{j=1}^{M_i} \exp\{\underline{\beta}'(\underline{x}_{ij}-\underline{x}_{i0})\}} \qquad (12)$$

(Breslow, Day, Halvorsen, Prentice and Sabai, 1978). If there is but a single control per case, this has precisely the form of the likelihood for ordinary logistic regression on the differences $\underline{x}_{i1}-\underline{x}_{i0}$, providing the constant term $\alpha$ is set to zero. Thus one may use conventional programs for logistic regression analysis to perform the conditional analysis for matched pairs (Holford, White and Kelsey, 1978).

## VII. CHOOSING BETWEEN CONDITIONAL AND UNCONDITIONAL LIKELIHOODS

In view of the computational problems which attend use of the conditional likelihood (11), one might consider that the unconditional model (7) is a better choice for general applications. Nevertheless it has serious drawbacks in precisely those situations where the conditional model is advantageous, namely when there are a large number of small strata. The problem, which is the familiar one of too many nuisance parameters, is perhaps best illustrated in the case of 1-1 pair matching with a single binary exposure variable.

Suppose there are I matched pairs consisting of a single case and a single control. The outcome for each pair may be represented in the form of a 2x2 table, of which there are four possible configurations

|          | E | Ē | E | Ē | E | Ē | E | Ē | Total (each table) |
|----------|---|---|---|---|---|---|---|---|--------------------|
| Case     | 0 | 1 | 1 | 0 | 0 | 1 | 1 | 0 | 1 |
| Control  | 0 | 1 | 0 | 1 | 1 | 0 | 1 | 0 | 1 |
| Total    | 0 | 2 | 1 | 1 | 1 | 1 | 2 | 0 | 2 , |

Number of such tables: (A)    (B)    (C)    (D)

where E denotes exposed and $\bar{E}$ non-exposed. According to the well-known theory of logistic or log-linear models (Fienberg, 1977; Haberman, 1974), unconditional maximum likelihood estimates are found by fitting frequencies to all cells in the three-dimensional 2x2xI configuration such that (i) the fitted frequencies satisfy the model; and (ii) their totals agree with the observed totals for each of the two-dimensional marginal tables. For the A concordant pairs in which neither case nor control is exposed, and the D concordant pairs in which both are exposed, the zero's in the margin require that the fitted frequencies be exactly as observed. The remaining B+C discordant pairs or tables have the same marginal configuration, namely one exposed and one non-exposed. For each such table the fitted frequencies are of the form

| | | |
|---|---|---|
| $\mu$ | $1-\mu$ | 1 |
| $1-\mu$ | $\mu$ | 1 |
| 1 | 1 | 2 |

where

$$\mu = pr_i(y=1|x=1) = \frac{\exp(\alpha_i+\beta)}{1+\exp(\alpha_i+\beta)}$$

and

$$1-\mu = pr_i(y=1|x=0) = \frac{\exp(\alpha_i)}{1+\exp(\alpha_i)} \quad ,$$

or in other words

$$\exp(\beta) = \Psi = \left(\frac{\mu}{1-\mu}\right)^2 .$$

The additional constraint satisfied by the fitted
frequencies is that the total number of exposed cases B+D
must equal the total of the fitted values (B+C)$\mu$+D, which
implies $\hat{\mu}$=B/(B+C). Consequently the unconditional maximum
likelihood estimate is

$$\hat{\Psi} = \left(\frac{\hat{\mu}}{1-\hat{\mu}}\right)^2 = \left(\frac{B}{C}\right)^2 .$$

Notice that one has also implicitly estimated the $\alpha_i$
nuisance parameters with this approach, such that
$\hat{\alpha}_i = \log\{(1-\hat{\mu})/\hat{\mu}\}$ for all pairs having non-zero marginals.

By way of contrast the contributions to the conditional
likelihood (11) are

$$\left(\frac{1}{1+\exp(-\beta)}\right)^B = \left(\frac{\Psi}{1+\Psi}\right)^B$$

for the B pairs in which the case is exposed and the control
not, and

$$\left(\frac{1}{1+\exp(\beta)}\right)^C = \left(\frac{1}{1+\Psi}\right)^C$$

for the C pairs with the control exposed and the case not,
the remaining (concordant) pairs contributing factors of
unity. Alternatively, one can verify directly that the
conditional distribution of B given B+C is binomial with
probability $\Psi/(1+\Psi)$. It follows that the conditional
maximum likelihood estimate is

$$\hat{\Psi} = \frac{B}{C} .$$

Hence the unconditional likelihood results in estimates of
the odds ratio which are the square of the correct,
conditional ones: a relative risk of 2 will tend to be
estimated as 4 by this approach, and that of 1/2 by 1/4.

Pike, Hill and Smith (1980) have investigated by
numerical means the asymptotic bias of the unconditional
maximum likelihood estimates of the odds ratio from matched
sets consisting of a fixed number of cases and controls.
For sets with 2 cases and 2 controls a true odds ratio of 2
tends to be estimated by values in range 2.51 to 2.53,
depending upon whether the exposure probability for controls
is .1 or .5. More generally, for sets containing a total of
N cases + controls, $\beta = \log(\Psi)$ tends to be estimated as $N\beta/$
$(N-1)$ (Breslow, 1981). Thus even with 10 cases and 10
controls per set an asymptotic bias of approximately 4%
remains when estimating a true odds ratio of 2. These
calculations demonstrate the need for considerable caution
in fitting unconditional logistic regression equations
containing many stratum or other nuisance parameters to
limited sets of data. Fortunately this is no longer a
problem now that methods based on the appropriate
conditional likelihood are becoming more generally
available.

Recognizing the problems caused by complex
stratification, Miettinen (1976) suggested computing a
"multivariate confounder score" as a means of incorporating
several confounding variables in the analysis yet keeping

the number of strata small.  He first discriminates cases
from controls by fitting a logistic equation or linear
discriminant which contains terms for the risk factor of
interest as well as for the confounders.  Setting the
coefficient of the risk factor to zero, the remainder of the
equation defines a score, on the basis of which each study
subject is assigned to one of five or so strata.  Estimation
of the adjusted relative risk then proceeds by standard
methods for stratified data analysis.  Unfortunately, while
this procedure yields relative risk estimates which are
reasonably free of bias, it can lead to a severe over-
statement of their statistical significance, and therefore
is not recommended (Pike, Anderson and Day, 1979).

## VIII.   RETROSPECTIVE REGRESSION MODELS

Since the independent probabilities in a case-control study
are those of exposures given disease status (9), rather than
vice-versa, some authors have found it more natural to model
these directly.  For situations having a single binary risk
variable x, and a vector of covariables $\underline{z}$ representing the
confounders, Prentice (1976) suggests fitting a
retrospective logistic model of the form

$$pr(x=1|y,\underline{z}) = \frac{\exp(\alpha+\beta y+\underline{\gamma}'\underline{z}+\underline{\delta}'y\underline{z})}{1+\exp(\alpha+\beta y+\underline{\gamma}'\underline{z}+\underline{\delta}'y\underline{z})} \ . \tag{13}$$

Here one adjusts for the confounders in terms of their
association with the exposure of interest, rather than their
relation to disease.  The analogous prospective model is

$$pr(y=1|x,\underline{z}) = \frac{exp(\alpha*+\beta x+\underline{\gamma}*\underline{z}+\underline{\delta}'x\underline{z})}{1+exp(\alpha*+\beta x+\underline{\gamma}*\underline{z}+\underline{\delta}'x\underline{z})} \quad , \quad (14)$$

where the notation is justified by the fact that both models
imply the same structure for the odds ratio relating x and
y, namely

$$\frac{pr(y=1|x=1,\underline{z})pr(y=0|x=0,\underline{z})}{pr(y=0|x=1,\underline{z})pr(y=1|x=0,\underline{z})}$$

$$= \frac{pr(x=1|y=1,\underline{z})pr(x=0|y=0,\underline{z})}{pr(x=0|y=1,\underline{z})pr(x=1|y=0,\underline{z})}$$

$$= exp\{\beta+\underline{\gamma}'\underline{z}\} \quad . \quad (15)$$

Inclusion of the interaction terms in (13) and (14) allows
modelling of changes in the relative risk according to the
covariables.

While the fitting of (13) yields perfectly adequate
estimates of the relative risk in this simple situation,
there are several reasons why (14) may be preferred. First,
it has the logical advantage of focussing the attention on
the correlates of disease, using the same model structure as
would be used to analyze prospective data from the same
population. Decisions about what covariates should be
included as confounders are often more easily made in terms
of their effect on disease rather than their relation to the
exposures (Breslow, 1982). One does not want to treat
factors which are intermediate in the putative causal chain
between exposure and disease as confounders, lest the true
association between exposure and disease be masked
(Miettinen, 1974). Another disadvantage of the
retrospective model is that it is awkward to generalize. If

a complex response variable $\underline{x}$, representing multiple
interacting risk factors, were used in (13), this would
require the fitting of large numbers of parameters. With
(14) the response variable is always binary, and the number
of parameters to be fitted is minimized. Finally, as the
modelling of the confounding effects becomes more flexible,
for example by the inclusion in $\underline{z}$ of higher order and
interaction terms involving the confounders, the two models
will yield progressively more similar estimates of the
relative risk parameters $\beta$ and $\underline{\delta}$ (Breslow and Powers, 1978).
This is true even when the exposures $\underline{x}$ are multiple and/or
continuous (Prentice and Pyke, 1979). Hence, from the
viewpoint of economy and interpretability, the prospective
model (14) seems clearly to be the model of choice, even
with case-control data.


IX.   EFFECT OF IGNORING THE MATCHING

Prior to the advent of methods for the multivariate analysis
of matched case-control studies, in particular those based
on the conditional likelihood (11), it was common practice
simply to ignore the matching in the analysis. In simple
problems one often found that accounting for the matching
made no perceptible change in the estimate of relative risk.
However, this was not always true and there was considerable
confusion regarding the conditions under which incorporation
of the matching in the analysis was necessary.

A sufficient and widely-quoted condition for poolability across matched sets or strata is that the stratification variables are either (i) conditionally independent of disease status given the risk factors, or (ii) conditionally independent of the risk factors given disease status. If either of these conditions is satisfied in the population from which the sample is drawn, both pooled and matched analyses provide asymptotically unbiassed estimates of the relative risk for a dichotomous exposure (Bishop, Fienberg and Holland, 1975). [Whittemore (1978) has shown that, contrary to previous belief, both types of analysis may sometimes yield equivalent results even if conditions (i) and (ii) are both violated.] In matched studies condition (i) is the more relevant since the matching variables are guaranteed to be uncorrelated with disease in the sample as a whole. Of course this does not ensure that they have the same distributions among cases and controls conditionally, within categories defined by the risk factors. While an unmatched analysis may therefore give biassed results, matching does guarantee at least that the bias will be conservative, i.e. that the relative risk estimate will be drawn towards unity (Seigel and Greenhouse, 1973; Armitage, 1975).

Generally, one must anticipate that the degree to which the matching variables are incorporated in the analysis will affect the estimates of relative risk. An example which well illustrates this phenomenon is provided by the joint

Iran/IARC study of esophageal cancer on the Caspian littoral
(Cook-Mozaffari, Azordegan, Day, Ressicaud, Sabai and
Aramesh, 1979). Here both cancer incidence and many
environmental variables show marked geographic variation and
cases and controls were therefore individually matched
according to village of residence as well as age. The data
were analyzed using both the conditional, fully matched
analysis based on (11) and the unconditional analysis based
on (9) where the entire sample was considered as a single
stratum. Intermediate between these two extremes additional
analyses were performed which incorporated various levels of
stratification by age and by geographic area, the latter
grouping the villages into regions with roughly homogeneous
incidence. Table 1 presents the results for males for four
risk variables which appeared to be the best indicators of
socio-economic and dietary status. Substantial bias of the
log relative risk estimates towards the origin, indicating a
lesser effect on risk, is evident with the coarsely
stratified and unmatched analyses. This confirms the
theoretical results noted above for the univariate
situation. While the standard errors of the estimates
increase with increasing matching, the changes are not great
and seem a small price to pay for the avoidance of bias.
Now that appropriate and flexible methods exist for doing
so, one should in fact always account for the matching in
the analysis.

Table 1  Estimates (± Standard Errors) of Log Relative Risk for Variables in the Multiple Regression Equation, Using a Variety of Analysis: Iran/IARC Esophageal Cancer Study

| Variables in equation | Fully matched | Type of analysis Stratified into: | | | | Unmatched |
| | | 7 Regions 4 age groups | 4 Regions 4 age groups | 4 Regions | 4 age groups | |
|---|---|---|---|---|---|---|
| Social class | -1.125±0.254 | -0.808±0.212 | -0.782±0.206 | -0.745±0.201 | -0.684±0.180 | -0.682±0.179 |
| Ownership of garden | -0.815±0.250 | -0.614±0.222 | -0.602±-.219 | -0.592±0.218 | -0.326±0.191 | -0.307±0.190 |
| Consumption of raw green vegetables | -0.552±0.220 | -0.459±0.203 | -0.439±0.199 | -0.432±0.198 | -0.429±0.188 | -0.440±0.187 |
| Consumption of cucumbers | -0.640±0.217 | -0.539±0.196 | -0.548±0.192 | -0.562±0.192 | -0.466±0.182 | -0.449±0.181 |
| Log likelihood | -187.69² | -388.27 | -388.80 | -390.40 | -393.53 | -394.78 |

1.  Source:  Breslow, Day, Halvorsen, Prentice, and Sabai (1978).
2.  Based on the conditional model and hence not comparable to the others.

X.  APPLICATIONS TO PARTICULAR DESIGNS

Using computer programs for the likelihood analysis of
conditional or unconditional (Baker and Nelder, 1978)
logistic regression models, the statistician or
epidemiologist may readily implement the methods of analysis
of case-control study data which are presented in this
chapter.  However, it is important to supplement the formal
regression analyses with tabulations of the basic data,
computations of relative risks in particular subgroups, and
plots of log relative risk with increasing dose, for
example, in order to have a better understanding for what is
actually going on.  Those who have long worked in this field
may feel that the simple methods of analysis developed for
grouped data are perfectly adequate if not even superior to
the computerized regression approach.  Some of their doubts
may be assuaged once it is realized that many of the
familiar and time-tested techniques are perfectly consistent
with the general logistic model.  In particular, several of
the classical tests for significance of the relative risk
can be viewed as "score" statistics (Rao, 1965) based on the
conditional likelihood (Day and Byar, 1979).  While
limitations of space prevent a full development of such
applications here, we nevertheless mention briefly two of
them.

A.  Combining Information from a Series of 2x2 Tables
The statistical problem which is perhaps most commonly
associated with case-control studies is that of estimating a

summary odds ratio from a series of 2x2 tables. Suppose
there are I such tables with entries denoted by the addition
of subscripts i to the frequencies (5), e.g. $a_i$ for the
number of exposed cases in the $i^{th}$ stratum, $n_{0i}$ for the
total number of controls. Cases and controls will have been
grouped into strata on the basis of covariables which are
thought either to confound or modify the effect of exposure
on disease, and we therefore assume that a p-vector $\underline{z}_i$ of
such covariables is associated with each table.

Following the general principles of inference outlined
above, we condition on the total set of exposures in each of
the i strata, which effectively means conditioning on the
marginal totals in each table. The likelihood (11) takes
the form of a product of non-central hypergeometric
densities

$$\prod_{i=1}^{I} \frac{\binom{n_{1i}}{a_i}\binom{n_{0i}}{m_{1i}-a_i}\Psi_i^{a_i}}{\sum_u \binom{n_{1i}}{u}\binom{n_{0i}}{m_{1i}-u}\Psi_i^u} , \qquad (16)$$

where $\Psi_i$ denotes the odds ratio in the $i^{th}$ table. Several
hypotheses about these odds ratios are of interest:

$$H_0: \quad \Psi_i = 1$$
$$H_1: \quad \Psi_i = \Psi = \exp(\beta)$$
$$H_2: \quad \Psi_i = \exp(\beta + \underline{\delta}'\underline{z}_i)$$
$$H_3: \quad \text{No restrictions on } \Psi_i.$$

$H_1$ expresses the notion that the relative risk of exposure
is constant across strata, while $H_2$ allows us to model
specific variations in risk associated with the covariables.

Gart (1971) considers the test of $H_0$ vs. $H_1$ and point
and interval estimates for $\Psi$ in $H_1$, while Zelen (1971)
develops exact tests for $H_2$ and $H_3$ vs. $H_1$. Breslow (1976)
discusses both conditional and unconditional estimates for
the parameters in $H_2$. While most of these procedures demand
iterative computation, even for this simple problem, the
score test of $H_0$ vs. $H_1$ does not. This takes the form

$$\chi^2 = \frac{\{\sum_{i=1}^{I} a_i - E_0(a_i)\}^2}{\sum_{i=1}^{I} \text{Var}_0(a_i)} \tag{17}$$

where $E_0(a_i) = n_{1i} m_{1i}/N_i$ and $\text{Var}_0(a_i) = n_{1i} n_{0i} m_{1i} m_{0i}/$
$\{N_i^2(N_i-1)\}$ are the exact conditional means and variances of
$a_i$ under $H_0$. This is the famous combined chi-square
statistic proposed by Cochran (1954) and (with the exact
variance) Mantel and Haenszel (1959). The extension of the
test to evaluate a trend in relative risk with increasing
dose (Mantel, 1963) likewise follows as the score test for
the appropriate model (Tarone and Gart, 1980).

No discussion of case-control studies would be complete
without mention of the robust estimate of $\Psi$ in $H_1$ proposed
by Mantel and Haenszel (1959):

$$\hat{\Psi}_{MH} = \frac{\sum_{i=1}^{I} a_i d_i/N_i}{\sum_{i=1}^{I} b_i c_i/N_i} \, .$$

This formula yields consistent estimates of $\Psi$ even in
situations where the strata are numerous and the data are

thin. For example, in matched pair studies, it is identical with the conditional maximum likelihood estimate B/C (see VII above). In view of its computational simplicity, consistency and efficiency (Breslow, 1981) this is clearly the estimate of choice for the non-specialist and it will undoubtedly continue to play an important role in the field. General purpose formulae for the variance of log $\hat{\Psi}_{MH}$ have recently been developed (Hauck, 1979, Breslow and Liang, 1982).

B.  1:M Matching with Dichotomous Exposure

If each case is matched to exactly M controls, the data may be summarized in the 2x(M+1) table

|  | 0 | 1 | 2 |  | M |
|---|---|---|---|---|---|
| Case exposed | $n_{1,0}$ | $n_{1,1}$ | $n_{1,2}$ | $\cdots$ | $n_{1,M}$ |
| Case not exposed | $n_{0,0}$ | $n_{0,1}$ | $n_{0,2}$ | $\cdots$ | $n_{0,M}$ |
|  | $T_1$ | $T_2$ | $T_3$ | | $T_M$ |

where entries denote frequencies of matched sets. Following general principles of conditional inference, these are grouped into pairs of sets for which the totals of exposed among cases and controls combined are equal. The likelihood (12) is proportional to

$$\prod_{m=1}^{M} \left( \frac{m\Psi}{m\Psi+M-m+1} \right)^{n_{1,m-1}} \left( \frac{M-m+1}{m\Psi+M-m+1} \right)^{n_{0,m}}, \quad (17)$$

an expression which may also be derived directly by noticing that the conditional distribution of $n_{1,m-1}$ given $T_m$ is binomial with probability $m\psi/(m\psi+M-m+1)$. Miettinen (1970) discusses in detail the estimation of $\psi$ from (17).

The Mantel-Haenszel estimate in this case may be written

$$\hat{\psi}_{MH} = \frac{\displaystyle\sum_{m=0}^{M} (M-m)n_{1,m}}{\displaystyle\sum_{m=0}^{M} m\, n_{0,m}},$$

and while different from the maximum likelihood estimate unless $M = 1$, nevertheless is consistent and reasonably efficient in comparison. A simple expression for the variance of $\log \hat{\psi}_{MH}$ (Connett, Ejigou, McHugh and Breslow, 1982) is

$$\hat{Var}(\log \hat{\psi}_{MH}) = \frac{\displaystyle\sum_{m=0}^{M} (M-m)^2 n_{1,m} + \hat{\psi}^2_{MH} \sum_{m=0}^{M} m^2 n_{0,m}}{\{\displaystyle\sum_{m=0}^{M} (M-m)n_{1,m}\}^2}.$$

REFERENCES

Anderson, J.A. (1972).  Separate sample logistic discrimination. *Biometrika*, 59: 19-35.

Armitage, P. (1975).  The use of the cross-ratio in aetiological surveys.  In *Perspectives in Probability and Statistics*, J. Gani (Ed.).  Academic Press, London, pp. 349-355.

Baker, R.J. and Nelder, J.A. (1978).  The GLIM System. Release 3.  Numerical Algorithms Group, Oxford.

Berkson, J. (1958). Smoking and lung cancer: Some observations on two recent reports. *Journal of the American Statistical Association*, 53: 28-38.

Bishop, Y.M.M., Fienberg, S.E., and Holland, P.W. (1975). *Discrete Multivariate Analysis: Theory and Practice.* MIT Press, Cambridge.

Breslow, N. (1976). Regression analysis of the log odds ratio: A method for retrospective studies. *Biometrics*, 32: 409-416.

Breslow, N. (1981). Odds ratio estimators when the data are sparse. *Biometrika*, 68:73-84.

Breslow, N. (1982). Design and analysis of case-control studies. *Annual Review of Public Health*, 3: 29-54.

Breslow, N.E. and Day, N.E. (1980). *Statistical Methods for Cancer Research 1: The Analysis of Case-Control Studies.* International Agency for Research on Cancer, Lyon.

Breslow, N.E., Day, N.E., Halvorsen, K., Prentice, R.L., and Sabai, C. (1978). Estimation of multiple relative risk functions in matched case-control studies. *American Journal of Epidemiology*, 108: 299-307.

Breslow, N.E. and Liang, K.Y. (1982). The variance of the Mantel-Haenszel estimator. *Biometrics*. In press.

Breslow, N. and Powers, W. (1978). Are there two logistic regressions for retrospective studies? *Biometrics*, 34: 100-105.

Cochran, W.G. (1954). Some methods for strengthening the common $\chi^2$ tests. *Biometrics*, 10: 417-451.

Connett, J., Ejigou, A., McHugh, R., and Breslow, N. (1982). The precision of the Mantel-Haenszel estimator in case-control studies with multiple matching. *American Journal of Epidemiology*. In press.

Cook-Mozaffari, P.J., Azordegan, F., Day, N.E., Ressicaud, A., Sabai, C., and Aramesh, B. (1979). Oesophageal cancer studies in the Caspian littoral of Iran: Results of a case-control study. *British Journal of Cancer*, 39: 293-309.

Cornfield, J. (1951). A method of estimating comparative rates from clinical data. Applications to cancer of the lung, breast and cervix. *Journal of the National Cancer Institute*, 11: 1269-1275.

Cornfield, J., Haenszel, W., Hammond, E.C., Lilienfeld, A.,
    Shimkin, M.B., and Wynder, E.L. (1959).  Smoking and
    lung cancer:  Recent evidence and a discussion of
    some questions.  *Journal of the National Cancer
    Institute*, *22*: 173-203.

Cox, D.R. (1970).  *The Analysis of Binary Data*.  Methuen,
    London.

Cox, D.R. (1972).  Regression models and life tables (with
    discussion).  *Journal of the Royal Statistical
    Society B*, *34*: 187-220.

Cox, D.R. (1975).  Partial likelihood.  *Biometrika*, *62*:
    599-607.

Cox, D.R. and Hinkley, D.V. (1974).  *Theoretical Statistics*.
    Chapman and Hall, London.

Day, N.E. and Byar, D. (1979).  Testing hypotheses in case-
    control studies:  Equivalence of Mantel-Haenszel
    statistics and logit score tests.  *Biometrics*, *35*:
    623-630.

Efron, B. (1975).  The efficiency of logistic regression
    compared to normal discriminant analysis.  *Journal
    of the American Statistical Association*, *70*:
    892-898.

Fienberg, S.E. (1977).  *The Analysis of Cross-Classified
    Categorical Data*.  MIT Press, Cambridge.

Gail, M.H., Lubin, J.H., and Rubinstein, L.V. (1981).
    Likelihood calculations for matched case-control
    studies and survival studies with tied death times.
    *Biometrika*, *68*: 703-707.

Gart, J.J. (1971).  The comparison of proportions:  A review
    of significance tests, confidence intervals and
    adjustments for stratification.  *International
    Statistical Review*, *39*: 148-169.

Gart, J.J. (1979).  Statistical analyses of the relative
    risk.  *Environmental Health Perspectives*, *32*:
    157-168.

Haberman, S. (1974).  *The Analysis of Frequency Data*.
    University of Chicago Press, Chicago.

Halperin, M., Blackwelder, W.C., and Verter, J.I. (1971).
    Estimation of the multivariate logistic risk
    function:  A comparison of the discriminant function
    and maximum likelihood approaches.  *Journal of
    Chronic Diseases*, *24*: 125-158.

Hauck, W.W. (1979). The large sample variance of the Mantel-Haenszel estimator of a common odds ratio. *Biometrics*, 35: 817-820.

Holford, T.R., White, C., and Kelsey, J.L. (1978). Multivariate analysis for matched case-control studies. *American Journal of Epidemiology*, 107: 245-256.

Liddel, D., McDonald, C., and Thomas, D. (1977). Methods of cohort analysis: Appraisal by application to asbestos mining (with discussion). *Journal of the Royal Statistical Society* A, 140: 469-491.

Mantel, N. (1963). Chi-square tests with one degree of freedom: extensions of the Mantel-Haenszel procedure. *Journal of the American Statistical Association*, 58: 690-700.

Mantel, N. (1973). Synthetic retrospective studies and related topics. *Biometrics*, 29: 479-486.

Mantel, N. and Haenszel, W. (1959). Statistical aspects of the analysis of data from retrospective studies of disease. *Journal of the National Cancer Institute*, 22: 719-748.

Miettinen, O. (1970). Estimation of relative risk from individually matched series. *Biometrics*, 26: 75-86.

Miettinen, O. (1974). Confounding and effect modification. *American Journal of Epidemiology*, 100: 350-353.

Miettinen, O.S. (1976). Stratification by a multivariate confounder score. *American Journal of Epidemiology*, 104: 609-720.

Pike, M.C., Anderson, J., and Day, N. (1979). Some insights into Miettinen's multivariate confounder score approach to case-control study analysis. *Journal of Epidemiology and Community Health*, 33: 104-106.

Pike, M.C., Hill, A.P., and Smith, P.G. (1980). Logistic analysis of stratified case-control studies. *International Journal of Epidemiology*, 9: 89-95.

Prentice, R.L. (1976). Use of the logistic model in retrospective studies. *Biometrics*, 32: 599-606.

Prentice, R.L. and Breslow, N.E. (1978). Retrospective studies and failure time models. *Biometrika*, 65: 153-158.

Prentice, R.L. and Pike, R. (1979). Logistic disease
    incidence models and case-control studies.
    *Biometrika, 66*: 403-411.

Press, S.J. and Wilson, S. (1978). Choosing between
    logistic regression and discriminant analysis.
    *Journal of the American Statistical Association, 73*:
    699-705.

Rao, C.R. (1965). *Linear Statistical Inference and its
    Applications*. Wiley, New York.

Seigel, D.G. and Greenhouse, S. W. (1973). Multiple
    relative risk functions in case-control studies.
    *American Journal of Epidemiology, 97*: 324-331.

Smith, P.G., Pike, M.C., Hill, A.P., Breslow, N.E., and Day,
    N.E. (1981). Multivariate conditional logistic
    analysis of stratum-matched case-control studies.
    *Applied Statistics, 30*:190-197

Tarone, R.E. and Gart, J.J. (1980). On the robustness of
    combined tests for trends in proportions. *Journal
    of the American Statistical Association, 75*:
    110-116.

Truett, J., Cornfield, J., and Kannel, W. (1967). A
    multivariate analysis of the risk of coronary heart
    disease in Framingham. *Journal of Chronic Diseases,
    20*: 511-524.

Whittemore, A.S. (1978). Collapsability of multidimensional
    contingency tables. *Journal of the Royal
    Statistical Society B, 40*: 328-340.

# 4
# METHODS OF SURVIVAL ANALYSIS

Dianne M. Finkelstein
Harvard University School of Public Health
and Sidney Farber Cancer Institute
Boston, Massachusetts

Robert A. Wolfe
University of Michigan
Ann Arbor, Michigan

## I. INTRODUCTION

In this chapter we review several data analytic methods which have proved valuable for medical research. These methods are useful for summarizing the lengths of times to an event of interest for a sample of subjects. Applications in cancer research have included the analysis of times to patient death, remission, or recurrence. Because the analysis of times to death is one of the most common applications of these methods, they are often call "survival" analysis methods, but they are by no means limited to the analysis of survival times.

127

The methods of survival analysis can lead to estimates and statistical inference for various parameters, including rates of occurrence of the event and median time to the event. In addition, methods are available for modelling the relationship between survival times and characteristics of the study subjects. For example, the relationship between patient age and cancer death rates could be modelled. As another example, the effects of differing treatment protocols on patient survival could be quantified. As a final example, the effects of treatment regimen on patient survival could be analyzed while simultaneously adjusting for patient age and other patient characteristics.

In a general framework, we will discuss methods which can be used to model and determine the association between certain covariates, X, and the distribution of the time, $S \geq 0$, to the occurrence of an event of interest. For data resulting from cancer research, S could be the time to death or the length of remission time for a subject. Such "failure time" data could arise from a clinical trial designed to compare two treatment programs with respect to time to recurrence of a disease, for example. Here, the covariate X would be an indicator of treatment group. Another example would be a longitudinal study in which an investigator is interested in the association between survival time and certain characteristics which have been recorded for each individual. It is of value, for example, to estimate the pattern of mortality among individuals with covariate values $X = x$, and more specifically to determine the conditional survival distribution $G(s|x) = P(S>s|x)$.

In most survival data, there are some failure times which are censored. Censored data occur when the time to occurrence of the event of interest is not known exactly and is only known to lie in some interval. Two types of censoring are commonly encountered. An observation is said to be right (left) censored when it is only known that the event occurred after (before) the observed time, T. Thus, we only know that the unobserved value of S is greater (less) than T. Data are said to be doubly censored if they have both left and right censored observations.

Right censored data are commonly found in longitudinal studies which end after a predetermined time. At the termination of the study, those individuals who have not yet had the event of interest will contribute right censored data. Double censoring can also arise in such follow-up studies (Gehan, 1965; Mantel, 1967; Peto, 1973). For example, if the event of interest is development of a tumor, then those individuals who have been first diagnosed as having the tumor in an early stage would be treated as if time to tumor were known exactly. Those with a widely metastasized, older growth would contribute left censored data. The healthy individuals would contribute right censored data. Another example is data which arise from animal sacrifice experiments. The analysis of early lesions of a lethal disease is appropriately done by treating the

data as left and right or interval censored (Hoel and
Walburg, 1972).

   In the case for which incomplete data consist of exact
times and right censored times, there are widely accepted
methods for determining the regression of S on X
(Kalbfleisch and Prentice, 1979).  However, these methods
are inappropriate for left censored data.  As a result, left
censored data are sometimes omitted or treated as exact,
which can bias the results of analysis (Hoel and Walburg,
1972).

   The development and refinement of survival analysis
methods is currently an area of active research with
numerous relevant publications in the statistical
literature.  A wide variety of survival analysis methods are
now well established and have been reviewed in several books
(Gross and Clark (1975); Kalbfleisch and Prentice, (1980)).
These methods have been given a solid theoretical framework
with the use of martingale theory as developed by Aalen
(1978), Gill (1980) and Anderson and Gill (1982).  We will
not give a comprehensive review of this survival analysis
literature, but will instead focus on survival analysis
methods appropriate for left and right censored data.  As
discussed above this is an important focus for cancer
research.

   The field of survival analysis offers a variety of
valuable methods and models for use on data that arise in
cancer research.  Several approaches, both parametric and

nonparametric, have been used in the analysis of such data.
The parametric methods are reviewed in the book by Gross and
Clark (1975) and in chapters two and three of the book by
Kalbfleisch and Prentice (1980). In this chapter, emphasis
is given to the nonparametric approaches. The origins of
such methods can be traced to work on the nonparametric
estimation of the distribution function for homogeneous
samples (Kaplan and Meier, 1958). Later, in the 1960's and
1970's, research moved towards the analysis of the
association between survival and concomitant variables (Cox,
1972; Kalbfleisch and Prentice, 1979; Miller, 1976). This
literature review is divided into two sections which reflect
this history of the field.


II.  ESTIMATION OF THE EMPIRICAL SURVIVAL DISTRIBUTION FROM
     INCOMPLETE DATA

One of the basic tools in the description of mortality
experience of a population , the life table, was developed
as early as 1693 by E. Halley in England. Later,
M. Greenwood (1926) gave formulas for the variance and
covariance of the life table functions. The estimates for
the life table functions are determined from survival time
data that are grouped into time intervals, and they are
therefore dependent on the choice of the time intervals. In
1958, Kaplan and Meier introduced the  product limit (P-L)
estimator which uses ordered observations instead of grouped
data. Herein lie the origins of modern survival analysis.

A.  Product Limit Estimator

In order to develop the Kaplan-Meier  product limit (P-L)
estimator, let the distinct times of death among the N
observed times be $s_1 \leq \ldots \leq \ldots \leq s_m$, and let $s_0 = 0$ and
$s_{m+1} = \infty$.  Let $d_j$ denote the number of deaths observed at
time $s_j$.  The estimator for $G(s) = P(S > s)$ is based on the
fact that at $s = s_k$, we can write

$$G(s_k) = P(S > s_k) = \prod_{j=1}^{k} P(S > s_j \mid S > s_{j-1}).$$

A natural estimator for $G(s_k)$ is thus given by

$$\hat{G}(s_k) = \prod_{j=1}^{k} \frac{n_j - d_j}{n_j}, \text{ for } k = 1, \ldots, m,$$

where $n_j$ is the number of individuals still at risk for
death just before $s_j$.

This survival distribution estimator has discrete steps
at each of the m failure times.  The P-L estimator maximizes
the likelihood:

$$L = \prod_{j=1}^{m} (G_{j-1} - G_j)^{d_j} G_j^{c_j}$$

where $G_j = G(s_j)$, $d_j$ is the number of deaths at $s_j$ and $c_j$ is
the number censored in the interval $[s_j, s_{j+1})$.  The
likelihood can be reparameterized and written as

$$L = \prod_{j=1}^{m} p_j^{n_j - d_j} q_j^{d_j} \tag{1}$$

where

$$p_j = 1 - q_j = P(S > s_j \mid S > s_{j-1}) = \frac{G(s_j)}{G(s_{j-1})}.$$

Each factor of (1) is maximized individually by the binomial estimate

$$\hat{p}_j = \frac{n_j - d_j}{n_j}.$$

Breslow and Crowley (1974) developed the asymptotic properties of this estimator.

Efron (1967) showed that the product limit estimate satisfies a condition he called "self-consistency":

$$N\hat{G}(s_k) = \sum_{j=k+1}^{m} d_j + \sum_{j=1}^{k-1} c_j \frac{\hat{G}(s_k)}{\hat{G}(s_j)} \text{ for } k = 1, \ldots, m \qquad (2)$$

where $\hat{G}(s_k)$ is the estimate for $G(s_k) = P(S > s_k)$

The consistency represented by this equation is that the expected number of survivors at time $s_k$ (the left side) is equal to the number who die at times after $s_k$ plus the number of those censored by $s_k$, who would be expected to survive beyond $s_k$ (the right side).

Efron described a method of constructing the unique estimate satisfying (2) for untied data. Construction begins by placing probability mass $1/N$ at each of the N observed times, $t_1, \ldots, t_N$ (censored or uncensored). Note that if there were no censoring, this would determine the empirical estimate of the survival curve. To account for the censoring, the masses at the censored times are now redistributed in the following way: If $t_j$ is the smallest time that is censored, remove the mass of the censored observation at $t_j$ and redistribute it equally among the $N-j$ times which are greater than $t_j$: $t_{j+1}, t_{j+2}, \ldots, t_N$. If

$t_\lambda$ is the smallest censored time among the remaining times,
$t_{j+1}$, $t_{j+2}$, ..., $t_N$, redistribute its mass, which will be
$1/N + 1/[N(N-j)]$ among the $N-\lambda$ times greater than $t_\lambda$: $t_{\lambda+1}$,
..., $t_N$. Continue in this fashion through the last censored
time. The resulting distribution will have mass at the
uncensored times, $t_k$ and at the last observed time (k=N),
censored or not. We note, therefore, that this "self-
consistent" estimate effectively treats the largest time
observation as uncensored and is otherwise the same as the
Kaplan-Meier point estimator. This treatment of the largest
observed time avoids many analysis problems that can arise
when the last observation is censored.

The values achieved by the self-consistent estimate for
the survival distribution, G, depend only on the order, or
ranks of the observed times and their censoring status but

Table 1   Ages-at-death of untreated RFM male mice dying
          with lung tumors

| Necropsy finding | Individual ages at death (days) |
|---|---|
| | Conventional mice (X = 1) |
| Lung tumor | 381, 539, 603, 616, 651 |
| No lung tumor | 479,500,516,553,572,594,632,642, 663, 673, 720, 760, 779 |
| | Germfree mice (X = 0) |
| Lung tumor | 692, 782, 846, 873, 888, 911, 921, 945, 1003 |
| No lung tumor | 817, 886, 942 |

not on the times themselves. Also, there is no assumption
made about the shape or form of G. The following example
illustrates the computation of the Kaplan-Meier estimate.
The data are a subset of the data from Hoel and Walburg
(1972).

The combined sample Kaplan-Meier estimate for survival for
these two groups is as follows:

| t | $\hat{S}(t)$ |
|---|---|
| 381 | .97 |
| 539 | .93 |
| 603 | .89 |
| 616 | .84 |
| 651 | .80 |
| 692 | .74 |
| 782 | .68 |
| 846 | .60 |
| 873 | .53 |
| 888 | .44 |
| 911 | .35 |
| 921 | .26 |
| 945 | .13 |
| 1003 | 0.00 |

B.  The Self-Consistent Estimate for Double-Censored Data

Turnbull (1974) extended the "self-consistent estimate" to
allow for left censored data. He showed that his estimate
can be constructed by an iterative procedure similar to the
one previously described by Efron, and showed how to
determine times $s_j$, $1 \le j \le m$, among the $t_i$ which could have
non-zero mass in the estimate. The procedure begins by
using an initial estimate for $G(s) = P(S>s)$ which is found
by ignoring the left censored observations and determining
the Kaplan-Meier estimate for the rest of the data. After
this, each step of this procedure uses a previous estimate,

$\hat{G}(s)$ to form an improved estimate $\hat{G}'(s)$ in the following
way: Let $d_j$ be the number of deaths, $c_j$ be the number of
right censored and $b_j$ be the number of left censored
observations at $s_j$, $1 \le j \le m$. The situation is illustrated by
the following tabulation:

| Type of Observation | Notation |
|---|---|
| Times | $\underline{s}_1, \underline{s}_2, \ldots, \underline{s}_m$ |
| Deaths | $d_1, d_2, \ldots, d_m$ |
| Left Censored | $b_1, b_2, \ldots, b_m$ |
| Right Censored | $c_1, c_2, \ldots, c_m$ |

The estimate $\hat{G}(s)$ is used to "adjust" the number of deaths
$d_j$ at each of the times $s_j$ for $j = 1, \ldots, m$ by adding a
"weight" $w_{ij}$ for each individual who is left censored and
for whom the observed time $s_i > s_j$. This weight is the
probability that the event occurred at time $s_j$ given that it
had occurred by $s_i$. Thus for these individuals,

$$w_{ij} = \frac{\hat{G}(s_{j-1}) - \hat{G}(s_j)}{1 - \hat{G}(s_i)}$$

From this, we obtain new, possibly non-integer values for
the expected number of deaths at $s_j$, for $j = 1, \ldots, m$,

$$d_j' = d_j + \sum_{i \ge j} b_i w_{ij} \tag{3a}$$

where we recall that $b_i$ is the number of left censored
individuals and $d_i$ is the number of deaths at time $s_i$.

   A new estimate $\hat{G}'(s)$ is now obtained by ignoring the
left censored observations and finding the product limit
estimate on the "adjusted" data set (using $c_j$ and $d_j'$ for

$j = 1,...,m$). Thus in addition to (3a), the following implicit equations are satisfied:

$$\hat{G}_1' = 1 - \frac{d_1'}{n_1'} ; \qquad \hat{G}_j' = \hat{p}_j \hat{G}_{j-1}'$$

$$\hat{G}_j' = \hat{p}(S > s_j), \qquad \hat{p}_j = \frac{n_j' - d_j'}{n_j'}; \qquad (3b)$$

$$n_j' = \sum_{i \geq j} (c_i + d_i').$$

This completes one iteration. This iterative procedure converges (Turnbull, 1974) and, in practice, the iterations are continued until a convergence criterion is satisfied.

Turnbull (1974) also showed that the set of implicit equations (3a and b) above satisfy the likelihood equations. Thus, the iterative procedure gives an estimate which maximizes the likelihood,

$$L = \prod_{j=1}^{m} (G_{j-1} - G_j)^{d_j} G_j^{c_j} (1 - G_j)^{b_j}.$$

In a 1976 paper, Turnbull extended the concept of self-consistency to arbitrarily grouped, truncated or censored data. For doubly censored data, the self-consistency condition becomes for each $j$:

$$N\hat{G}_j = \sum_{i>j} d_i + \sum_{i \geq j} b_i \frac{\hat{G}(s_j)}{1 - \hat{G}(s_i)} + \sum_{i<j} c_i \frac{\hat{G}(s_j)}{\hat{G}(s_i)} \qquad (4)$$

where $d_i$ is the number uncensored at $s_i$, $b_i$ is the number left censored at $s_i$, and $c_i$ is the number right censored at $s_i$. The estimates $\hat{G}(s)$ which satisfy the self-consistency condition (4) are the maximum likelihood estimates (Turnbull, 1976). If there is no left censoring, $\hat{G}(s)$ will

coincide with the Kaplan-Meier estimates.  If there are no
uncensored observations (only left and right censored), this
estimate coincides with the Ayer $et\ al.$ (1954) estimate of
the empirical distribution function.  The Ayer estimate
maximizes the likelihood,

$$L = \prod_{i=1}^{m} G(s_i)^{c_i}(1-G(s_i))^{b_i},$$

subject to the constraint that $G(s_j)$ is monotonically non-
increasing.  Ayer $et\ al.$ showed that this estimate can be
obtained by the following procedure:

1)  If the sequence $\left\{ \dfrac{c_j}{b_j + c_j} \right\}_{j=1}^{m}$

   is non-increasing then the estimate for $G(s_j)$ will

   be $\dfrac{c_j}{c_j + b_j}$ for $j = 1,\ldots,m$.

2)  If $\dfrac{c_j}{c_j + b_j} < \dfrac{c_{j+1}}{c_{j+1} + b_{j+1}}$ for some $j$,

   then the estimate for $G(s_j)$ and $G(s_{j+1})$ will be

   $$\dfrac{c_j + c_{j+1}}{(c_j+c_{j+1}) + (b_j+b_{j+1})}$$

This procedure is repeated until a monotonic sequence is
obtained.  The method of grouping the elements of a sequence
to obtain an estimate under order restrictions was further
developed by Barlow, Bartholomew, Bremner and Brunk (1972)
into the concepts of isotonic regression.

   For an example of the Ayer estimate, we consider again
the data of Hoel and Walburg (1972) given in Table 1.
Often, it is appropriate to treat time to tumor data as

right and left censored rather than as right censored and
exact due to the uncertainty surrounding the time at which
the tumor began.  Thus the more appropriate estimate for the
survival curve is the Ayer estimate rather than the Kaplan-
Meier estimate.  For the data given in Table 1, the Ayer
estimate for the survival distribution for the combined
sample is as follows:

| s | $\hat{G}(s)$ |
|-----|------|
| 381 | .75 |
| 603 | .64 |
| 782 | .50 |
| 846 | .33 |
| 888 | .25 |
| 945 | 0 |

Turnbull's self-consistent estimate is a generalization
of both the Kaplan-Meier and the Ayer estimates.  It is a
non-parametric estimate for a survival distribution function
and leads to a decreasing step function.  Turnbull, in a
lemma in his 1976 paper, determined where the steps occur:

1)  For right censored and uncensored data, the steps
    (in the Kaplan-Meier estimate) will be at the
    uncensored times.

2)  For left censored and right censored data, the
    steps (in the Ayer estimate) will be at left
    censored times.

3)  For the doubly censored data, the steps (in
    Turnbull estimate) will be at the uncensored times
    and at left censored times whose immediate
    predecessor was a right censored time.

The estimators discussed so far have been for the
empirical survival function G(s) of a homogeneous
population. Recent interest has centered on comparison of
survival curves and more generally on determining estimates
for the conditional survival distribution, G(s|x). We now
discuss these methods.


III. NONPARAMETRIC METHODS FOR THE COMPARISON OF SURVIVAL
     DISTRIBUTIONS


A theoretical development of a variety of statistical
testing procedures for the comparison of survival
distributions was given by Anderson *et al*. (1982). Here, we
develop some of the classical tests that have been used for
such purposes.


A. Comparison of Two Samples

The problem of comparing survival distributions often
arises in cancer studies. It may be of interest, for
example, to compare the ability of two treatments to prolong
life. If the data were uncensored, it would be appropriate
to use the Wilcoxon Test for comparison of the two
distributions. We develop generalizations for the Wilcoxon
Test because of its widespread use and its historical
importance, although today the logrank test is more commonly
used than the generalized Wilcoxon. To develop the

Wilcoxon, let the failure times from the two treatment groups be $X_{11}, \ldots, X_{1m}$; $X_{21}, \ldots, X_{2n}$. For computational simplicity, it is desirable to develop the Mann-Whitney form of the Wilcoxon Test. For this test, define

$$U(X_{1i}, X_{2j}) = U_{ij} = \begin{cases} 1 & \text{if } X_{1i} > X_{2j} \\ 0 & \text{if } X_{1i} = X_{2j} \\ -1 & \text{if } X_{1i} < X_{2j} \end{cases}$$

$$\text{and } U = \sum_{i=1}^{m} \sum_{j=1}^{n} U_{ij}.$$

The null hypothesis of equality of the two distributions should be rejected for very large or very small U. In fact, the test statistic used is

$$Z = \frac{U - E(U)}{\sqrt{Var(U)}} = \frac{U}{\dfrac{\sqrt{mn(m+n+1)}}{\sqrt{3}}}$$

where $E(U)$ and $Var(U)$ are the moments calculated under the null hypothesis. The test statistic Z is asymptotically normal with mean 0 and variance 1. Gehan (1965) gave a generalization of this test which can be used with censored data. Mantel's (1967) computational form for this test is easier to work with and is as follows. Let the combined sample be denoted $(Z_1, \delta_1; \ldots; Z_{m+n}, \delta_{m+n})$ where $\delta_i$ is the indicator for censoring ($\delta_i = 0$ if $Z_i$ is right censored).

Let

$$
U_{kj} = U(Z_k, Z_j) = \begin{cases} 1 \text{ if } (Z_k > Z_j, \; \delta_j = 1) \text{ or} \\ \qquad (Z_k = Z_j, \; \delta_k = 0, \; \delta_j = 1) \\ -1 \text{ if } (Z_k < Z_j, \; \delta_k = 1) \text{ or} \\ \qquad (Z_k = Z_j, \; \delta_k = 1, \; \delta_j = 0) \\ 0 \text{ otherwise,} \end{cases}
$$

$$
U_k' = \sum_{\substack{j=1 \\ j \neq k}}^{m+n} U_{kj}
$$

$$
U = \sum_{k=1}^{m+n} U_k'
$$
Sample 1

The variance of U under a permutation model in which we consider that the m observations from Sample 1 were drawn from the urn containing $(Z_1, \ldots, Z_{m+n})$ is, under the null hypothesis,

$$
\text{Var}(U) = \frac{mn}{m+n(m+n-1)} \sum_{i=1}^{m+n} (U_i')^2
$$

Under $H_o$, $U/[\text{Var}(U)]^{1/2}$ is asymptotically normal with mean 0 and variance 1. For the data given in Table 1, if we treat the observations as right censored and exact, Z = 2.02 and the hypothesis of equality of the survival distributions marginally is rejected (p = 0.04). One concludes that survival is better for the conventional mice.

This test has been extended by Gehan so that it is appropriate for doubly censored data (Gehan, 1965 b). In order to describe such data, define the three indicators:

$\delta_i$, for uncensored; $\lambda_i$, for right censored; and $\mu_i$, for left censored. Thus, the two samples are

$\{(X_{11}, \delta_{11}, \lambda_{11}, \mu_{11}), \ldots, (X_{1m}, \delta_{1m}, \lambda_{1m}, \mu_{1m})\}$ and

$\{(X_{21}, \delta_{21}, \lambda_{21}, \mu_{21}), \ldots, X_{2n}, \delta_{2n}, \lambda_{2n}, \mu_{2n})\}$. Let:

$$
U_{ij} = \begin{cases}
-1 \text{ if } X_{1i} < X_{2j} \text{ and } \delta_{1i} = \delta_{2j} = 1 \text{ or} \\
\qquad X_{1i} \leq X_{2j} \text{ and } \delta_{1i} = 1, \lambda_{2j} = 1 \text{ or} \\
\qquad X_{1i} \leq X_{2j} \text{ and } \mu_{1i} = 1 \text{ and } \mu_{2j} = 0. \\
+1 \text{ if } X_{1i} > X_{2j} \text{ and } \delta_{1i} = \delta_{2j} = 1 \text{ or} \\
\qquad X_{1i} \geq X_{2j} \text{ and } \lambda_{1i} = 1, \delta_{2j} = 1 \text{ or} \\
\qquad X_{1i} \geq X_{2j} \text{ and } \mu_{1i} = 0, \mu_{2j} = 1 \\
0 \quad \text{Otherwise.}
\end{cases}
$$

Calculate the statistic

$$
U = \sum_{i=1}^{m} \sum_{j=1}^{n} U_{ij}.
$$

The test statistic $Z = [U-E(U)]/[Var(U)]^{1/2}$ is asymptotically $N(0,1)$. For the variance, the following notation is necessary. Let:

$d_j$ = number of uncensored observations at rank j in the rank ordering of uncensored observations with distinct values, $t_1, \ldots, t_p$.

$l_j$ = number of right censored observations with value greater than observations at rank j but less than observations at rank j+1, for j=1,\ldots,p.

$m_j$ = number of left censored observations with value less than observations at rank j but greater than observations at rank j-1, for j=1,\ldots,p.

$$
D_k = \sum_{j=1}^{k} d_j, \quad D_0=0 \quad L_k = \sum_{j=1}^{k} l_j, \quad L_0=0 \quad M_k = \sum_{j=1}^{k} m_j, \quad M_0=0.
$$

The variance under the null hypothesis of equality of the survival distributions is given by

$$
\begin{aligned}
\mathrm{Var}(U) = & \frac{mn}{(m+n)(m+n-1)} \left\{ \sum_{j=1}^{p} d_j (D_{j-1}+M_j)(D_{j-1}+M_j+1) \right. \\
& + \sum_{j=1}^{p} l_j (D_j+M_j)(D_j+M_j+1) \\
& + \sum_{j=1}^{p} d_j (D_p+L_p-D_j-L_{j-1})(D_p+L_p-3D_{j-1}-L_{j-1}-2M_j-d_j-1) \\
& \left. + \sum_{j=1}^{p} m_j (D_p+L_p-D_{j-1}-L_{j-1})(D_p+L_p-D_{j-1}-L_{j-1}-1) \right\}.
\end{aligned}
$$

For the data given in Table 1, if we treat the observations as right and left censored, $Z = 0.67$ and one concludes that survival is better for the germfree mice, but the differences are not statistically significant.

The Gehan generalization to the Wilcoxon test has been the subject of recent criticism (Prentice and Marek, 1979) due to the fact that with the presence of censored data, it tends to inappropriately give too much weight to early failures. An alternative generalization of the Wilcoxon statistic is the Peto and Peto (1972) test. This test assigns a score to each observation which is a function of the survival distribution, $F(t_j)$, $j=1,\ldots,p$. For an uncensored observation $t_i$, the score is $u_i = F(t_i)-(1-F(t_{i-1}))$. For a censored observation $t_i^{+}$, the score is $u_i = F(t_i)-1$. These scores sum identically to 0. The test is based on the sum $U$ of the $U_i$ scores from one of the groups (call it the treatment group).  The variance of U is

$$V = \frac{mn}{(m+n)(m+n-1)} \sum_{i=1}^{m+n} U_i^2.$$

The statistics for the Peto and the Gehan tests can be written in a generalized form which allows comparison:

$$U = \sum_{j=1}^{p} w_j (z_j - \frac{m_j}{m_j + n_j}),$$

where $m_j$ and $n_j$ are the number at risk in each group at time $t_j$ and $z_j$ is the number of failures for the treatment group at time $t_j$, for $j=1,\ldots,p$.  For the Peto test, $w_j$ corresponds to $F_j$, the Kaplan-Meier estimate of survival at $t_j$.  For the Gehan test, $w_j = n_j$.  If a weight of $w_j = 1$ is used, for $j=1,\ldots,p$, this statistic becomes the logrank test.  For uncensored data, the first two of these tests are just the Wilcoxon test.  However, for censored data, it has been shown that the Peto weights are more appropriate.

B.  Comparison of K Samples

Comparison of K samples can be made using the Kruskal-Wallis nonparametric test if the data are uncensored.  For censored data, a generalization of this test has been proposed by Breslow (1970).  To test the null hypothesis, that the K samples are from the same survival distribution, we calculate scores,

$$U_i = \sum_{i \neq j}^{N} U_{ij}$$

for all of the N observations by any of the methods given

above.  Let $S_k$ be the sum of the scores $U_i$ for the $k^{th}$
sample, and $n_k$ the sample size for the $k^{th}$ group, $N=\Sigma n_k$.
The test statistic,

$$\chi^2 = \frac{\sum\limits_{k=1}^{K} S_k^2/n_k}{\sum\limits_{i=1}^{N} U_i^2/(N-1)}$$

is $\chi^2$ distributed asymptotically with K-1 degrees of freedom
(Peto and Peto, 1972).  Statistics based upon matrix
calculations are more appropriate whenever a computer can be
used (Andersen et al. (1982).


IV.  ANALYSIS OF SURVIVAL DATA WITH CONCOMITANT VARIABLES

During the 1960's, most attention was focused on introducing
covariates into parametric models for survival which would
be appropriate for (right) censored data.  Fiegl and Zelen
(1965) introduced a model in which survival time is assumed
to be exponentially distributed.  In this model, the mean
survival time for the $i^{th}$ individual in the study is assumed
to be a linear function of the concomitant variable $X_i$.
Thus the model is

$$P(S=s_i) = \theta_i e^{-\theta_i s_i} \text{ where } \theta_i^{-1} = a + bx_i.$$

Zippen and Armitage (1966) extended the results of Fiegl and
Zelen to censored data.  Glasser (1967) gave a log-linear
exponential model in which a constant hazard rate is assumed
and time is treated as a covariate.

## V.  COX PROPORTIONAL HAZARD MODEL

In 1972, D.R. Cox presented a very innovative paper before
the Royal Statistical Society in which he introduce the
proportional hazards model for survival S as a function of
covariate $\underline{z}$.  The model is a nonparametric generalization of
Glasser's (1967) model.  A unique aspect of Cox's model is
that rather than modeling the relationship between the
survival distributions and covariates, it models the
relationship of the hazard function and the covariates.

The hazard function, or instantaneous failure rate is
(Gehan, 1969)

$$\lambda(s) = \lim_{\Delta s \to 0} \frac{P(s<T<s+\Delta s \mid T>s)}{\Delta s} = -\frac{G'(s)}{G(s)}.$$

This is the instantaneous probability of death at time s for
those who were alive at time s.  In Cox's model, it is
assumed that covariates $\underline{z}$ act multiplicatively on the
underlying hazard function, $\lambda_0(s)$ so

$$\lambda(s;\ z) = \lambda_0(s)e^{\underline{z}\beta}. \tag{5}$$

The factor $e^{\underline{z}\beta}$ is, therefore, the instantaneous risk of
failure for individuals with covariates z relative to that
for individuals with z=0 (Instantaneous relative risk).

The Cox model is a semi non-parametric model since it
is possible to estimate the parameter $\beta$ without restricting
the shape of $\lambda_0(\cdot)$.  This is due to the fact that in Cox's
approach, $\beta$ is determined by maximizing a "partial
likelihood" (Cox, 1975) which is obtained in the following

way.   Let the data have no tied failure times.  For a
failure at time $s_j$, conditional on the risk set $R(s_j)$ (the
set of indices for individuals observed alive just before
time $s_j$), the probability that the failure is for the
individual with covariates equal to $\underline{z}_j$ is.

$$\frac{\lambda_0(s)e^{z_j\beta}}{\sum_{\ell\varepsilon R(s_j)}\lambda_0(s)e^{z_\ell\beta}} = \frac{e^{z_j\beta}}{\sum_{\ell\varepsilon R(s_j)}e^{z_\ell\beta}}.$$

Each of the m failures contribute a factor of this type to
the likelihood which becomes

$$L = \prod_{j=1}^{m} \{ \frac{e^{z_j\beta}}{\sum_{\ell\varepsilon R(s_j)}e^{z_\ell\beta}} \}. \tag{6}$$

Cox (1975) indicated that under certain mild
conditions, the partial likelihood (6) could be used for
asymptotic inference on $\beta$.  Kalbfleisch and Prentice (1973)
showed that in the absence of censoring, (6) is the
likelihood based on the marginal distribution of the ranks
of the failure times.  Breslow (1974) derived (6) as a
"maximized" likelihood for $\beta$ by restricting $\lambda_0$ to be a step
function with discontinuities at failure times.  Tsiatis
(1981) first gave strong consistency and asymptotic
normality for the Cox regression model.  Andersen and Gill
(1982) have established the consistency and asymptotic
normality of $\hat{\beta}$ under more general conditions.

Several generalizations of (6) have been proposed to
accommodate tied failure times (in which case S can no

longer be treated as a continuous random variable)
(Kalbfleisch and Prentice, 1973; Breslow, 1974).  If there
is only a small fraction of tied failures, then the various
generalizations are all well approximated by:

$$L = \prod_{j=1}^{m} \left\{ \frac{e^{v_j \beta}}{\left| \sum_{\ell \varepsilon R(s_j)} e^{z_\ell \beta} \right|^{d_j}} \right\} \qquad (7)$$

where $d_j$ is the number of failures at time $s_j$ and $v_j$ is the
sum of the covariate values for the $d_j$ individuals who have
failed at times $s_j$ (Kalbfleisch and Prentice, 1980).

Standard asymptotic procedures applied to (6) or (7)
give estimates for $\beta$ and tests of hypothesis about $\beta$.  At
$\beta = 0$, the score statistic from (7) is

$$U = \left. \frac{\partial \log L}{\partial \beta} \right|_{\beta = 0} = \sum_{i=1}^{m} [v_i - d_i n_i^{-1} \sum_{\ell \varepsilon R(s_i)} z_\ell]$$

where $U$ and $z_\ell$ may be vector valued and where $n_i$ is the
number of individuals in $R(s_i)$.  The Mantel-Haenszel
(Mantel, 1966) or log rank (Peto, 1972) test for the
comparison of several curves is a special case of the scores
test for this regression model with dummy covariates $z$.

When the number of tied failure times is too large, the
approximation given by (7) may be poor and a discrete
failure time model is preferred.  Cox (1972) suggested a
generalization of the partial likelihood method, in which
the conditional probability of failure was modeled as a
logistic regression model so

$$\frac{\lambda(s, z) \, dt}{1-\lambda(s, z)dt} = e^{z\beta} \cdot \frac{\lambda_0(s)dt}{1-\lambda_0(s)dt}$$

Here, $\beta$ is the log odds ratio for death. Alternatively, Kalbfleisch and Prentice (1973) showed that grouping the data in the continuous model (5) gives a discrete model which retains the log relative risk parameter $\beta$. Prentice and Glocker (1978) gave asymptotic likelihood methods based on this grouped failure model.

For the data given in Table 1, if we let group membership be the covariate z for the Cox model, and treat the data as right censored and exact, we obtain the estimate for $\beta$ of $\hat{\beta} = 2.05$. A test of the hypothesis H: $\beta=0$ tests whether the hazards are the same for the two groups. The result of the likelihood ratio test on the data from Table 1 is $\chi^2 = 4.6$. Thus we reject the hypothesis that the hazards for the two groups are equal (p = .035).

Much of the work in survival analysis during the 1970's was directed toward application and extension of Cox's model. There has been a generalization to models with competing risks of failure. (Gail, 1975 gives a review of the literature on competing risk.) Another direction of study has been towards analysis using models which allow for time-dependent covariates (Cox, 1972). In the model (5), only one value of Z (usually the study baseline value) could enter for each individual. In the model which allows covariates to change, Z is replaced by Z(s).

Other models have been presented which allow for changes in the covariates. Woodbury *et al.* (1979) used a

model which incorporates both the density function for the
risk variables and the exponential survival function:

$$e^{-\mu(x_{is})} = P(T>s+1|T>s, x_{is})$$

for each time interval separately. For this model, $x_{is}$ is
the value of the covariate x at time s for individual i and
s+1 is the next time measurement after s. Thus, the
likelihood becomes

$$L = \prod_{s} [\prod_{\substack{\text{survivors} \\ \text{in } (s, s+1)}} \phi_s(x_{is})e^{-\mu(x_{is})}] \cdot$$

$$[\prod_{\substack{\text{deaths} \\ \text{in} \\ (s, s+1)}} \phi_s(x_{is})(1-e^{-\mu(x_{is})})]$$

(8)

where $\phi_s(x_{is})$ is the density function for the risk variable
x for individual i at time s. Since the risk variable
density function can be factored out, it is possible to
estimate parameters of the risk function separately from the
parameters of the function used to predict future risk
variables values. Further, the model allows maximal use of
incomplete data since the information from each individual
need only contribute to the likelihood for the time
intervals in which complete information is available.

We note that a common feature of the likelihood given
in (8) above and others discussed in this section is that
this likelihood has been factored into a product over time
intervals between uncensored times. These factors are
probabilities which are conditional on a risk set. It is
therefore only possible to use the information from
individuals who are known to be alive or whose exact death

times are known.  Thus, there is no simple way to include
left censored data into these models, and it is not uncommon
to find in applications that left censored individuals are
either omitted or incorrectly entered as uncensored events
(Hoel and Walburg, 1972).

B.  Regression Models Which Allow for Doubly Censored Data

One of the recently developed methodologies for dealing with
doubly censored data is an extension of regression
techniques.  Miller (1976) noted that if the data are
uncensored, then analysis can be done by a least squares
regression of survival time S (or a linearizing function of
the survival times) on the covariates X.  The model would be
written

$$s_i = a + bx_i + \varepsilon_i, \qquad i = 1,\ldots,N.$$

The least squares estimate $\hat{a}$ and $\hat{b}$ are those values of a and
b which minimize the sum of squares $\Sigma(s_i - a - bx_i)^2$, or
equivalently which minimize

$$(1/N)\Sigma(s_i - a - bx_i)^2 = \int \varepsilon^2 d\hat{F}_{ab}(\varepsilon) \qquad (9)$$

where $\hat{F}_{ab}(\varepsilon)$ is the sample distribution estimate based on
the residuals $\varepsilon_i = s_i - a - bx_i$ for $i = 1,\ldots,N$.

Miller suggested that an extension of regression to
right censored data can be done by choosing values of a and
b which again minimize

$$\int \varepsilon^2 d\hat{F}_{ab}(\varepsilon)$$

where, in the case of right censored data, $\hat{F}_{ab}(\varepsilon)$
is the Product-Limit estimate based on censored and

uncensored residuals $\varepsilon_i(ab) = s_i-a-bx_i$. In the case of doubly censored data, the Turnbull estimate is used. Hence, $\hat{F}_{ab}(\varepsilon)$ is discrete, taking jumps at uncensored or left censored points only. The least squares estimators obtained from this method are both computationally and theoretically difficult. Therefore, Miller suggested a modified estimator. The suggestion is to obtain an initial estimate $\hat{b}_0$ from a regression on only the uncensored observations. The residuals determined by this regression are now used to assign a weight $w_i(\hat{b}_0)$ to the $i^{th}$ observation, for $i = 1,\ldots,N$. The weight is determined by the jumps in the Kaplan-Meier estimate for the distribution of the residuals. The "modified" estimates $\hat{a}$ and $\hat{b}$ are now chosen to be those values which minimize the weighted sum of squares,

$$\Sigma_{uc}w_i(\hat{b}_0)(s_i-a-bx_i)^2$$

where the sum is over the uncensored observations only, since $w_i(\hat{b}_0)$ is 0 for censored observations. Thus, the estimate that results satisfies:

$$\hat{b} = \Sigma_{uc}w_i(\hat{b}_0)y_i(x_i-\hat{x})/\{\Sigma_{uc}w_i(\hat{b}_0)(x_i-\hat{x})^2\} \qquad (10)$$

where $w_i(\hat{b}_0)$ is the probability mass assigned by $\hat{F}_{ab}(\varepsilon)$ to uncensored residuals $\varepsilon_i(a, b)$ (a mass which does not depend on a), and $\hat{x} = \Sigma_i w_i(\hat{b}_0)x_i$. It is possible for equation (10) to yield zero, one, or multiple solutions. If only one solution were found to (10), there would be an obvious temptation to obtain an improved estimate by substituting $\hat{b}$ for $\hat{b}_0$ and iterating until convergence occurs. However, as

Miller notes, the estimate may not converge and in fact may oscillate, because small changes in the slope may change the relative ordering of the residuals.  This, in turn, may cause a discontinuous change in the weights.

The unstable nature of Miller's estimators detracts from the general usefulness of the method.  Further, we note that this method does not utilize the covariate information from censored observations.

Buckley and James (1979) modified Miller's approach to the estimation of the regression parameters.  Rather than altering the sum of the residuals, they modified the normal equations.  Thus, they noted that with uncensored data, they could choose estimates, $\hat{a}$ and $\hat{b}$ whose values satisfy

$$\sum_{i=1}^{N} (s_i - a - bx_i) = 0; \quad \sum_{i=1}^{N} (x_i - \hat{x})(s_i - bx_i) = 0.$$

To allow for censoring, the authors replaced $s_i$ in the normal equations by the expected value of $s_i$, $s_i^* = E(S_i | t_i, \delta_i)$ where $\delta_i$ is an indicator on censoring. Since $E(S_i | t_i)$ is unknown, a "self-consistent" approach is suggested to estimate it from the  product limit estimator, $\hat{F}_{a,b}(\epsilon)$ (or for doubly censored data, the Turnbull estimator).  For censored individuals, we expect that the residual is larger than that which has been observed.  This increase will be a weighted sum of all larger residuals (since it is known that the true death time S is larger than the observed time $(T_i = t_i)$).  Thus, the expected value, $E(S_i | T_i = t_i)$ will be

$$s_i^*(b) = bx_i + \sum_{uc} {}_k v_{ik}(b)(s_k - bx_k)$$

where the weights $v_{ik}$ are assigned according to the estimated probability that the true $i^{th}$ residual equals
$(s_k - bx_k)$:

$$v_{ik}(b) = \frac{w_k(b)}{1 - \hat{F}_{0,b}(t_i - bx_i)} \text{ for residuals } e_i(0,b) < e_k(0,b),$$

$$0 \text{ otherwise.}$$

Solving the normal equations leads to a new estimate
for b. By making successive substitutions, we wish to find
an estimate $\check{b}$ which satisfies the self-consistency condition

$$\check{b} = \frac{\sum_{uc} s_i(x_i - \bar{x}) + \sum_c s_i^*(\check{b})(x_i - \bar{x})}{\sum_{i=1}^{N} (x_i - \bar{x})^2}.$$

However, as was found with Miller's method, this procedure
does not always yield a unique solution. Further work is
required on the conditions necessary for convergence and the
asymptotic properties for both the Miller and the Buckley
and James estimators. Koul $et$ $al.$ (1981) have recently
given a modification of the Buckley-James method which leads
to consistent, asymptotically normal estimators.

Buckley's method uses an iterative procedure consisting
of two steps. In the first step, we estimate the expected
value of observations, conditional on the previous estimate
of the parameters and the observed (possibly censored)
values. In the second step, these expected values are
substituted into the normal equations to determine a new

estimate for the parameters. The two step procedure is
repeated until a stable solution (solutions) is achieved.

A general treatment of the iterative technique for
determining maximize likelihood estimates from incomplete
data (the  E-M Algorithm) is given in a paper by Dempster $et$
$al.$ (1977). Since censored data is incomplete, this is an
important contribution to the field of survival analysis.


V.  E-M ALGORITHM

The E-M Algorithm (EM) is an iterative procedure for finding
the maximum likelihood estimate from incomplete data.

The concept "incomplete data" implies the existence of
an "unobserved sample", x, the complete data which are
unobserved.  Additionally, there are two likelihoods:
$f(x|\phi)$, the likelihood for the complete data given the
parameter $\phi$, and $g(y|\phi)$, the likelihood for the incomplete
data, y, given the parameter $\phi$. The goal is to determine
the value of $\phi$ which maximizes log $f(x|\phi)$. However, log
$f(x|\phi)$ is unknown. Therefore, we maximize the conditional
expected value of it given the incomplete data, y, and the
current estimate for $\phi$, $\phi*$.

Each stage of the algorithm uses the previous estimate
of $\phi$, $\phi*$ to compute a new estimate $\phi**$. Each stage consists
of two steps:  the E step and the M step. In the E step, we
find the conditional expectation:

$E[\ln f(x|\phi)|y, \phi*] = Q(\phi|\phi*)$.

In the M step, we determine that value of $\phi$, $\phi**$ which
maximizes $Q(\phi|\phi*)$.

Dempster $et$ $al.$ (1977) showed that the likelihood Q
increases with each iteration (stage) of the algorithm.  The
convergence of the EM algorithm has been examined and
conditions under which it converges are presented in a
recent publication of Wu (1982).  The algorithm will not
always converge to the maximum likelihood estimate.  An
exception could occur if the iterative procedure converged
to a solution which was in fact a saddlepoint or a relative
maximum of the likelihood.  Determination of whether a point
is a maximum can be accomplished by a search procedure.

When the density $f(x|\phi)$ is from an exponential family,
the EM is simplified.  The expectation, or E step, consists
of finding the expected value of the sufficient statistic,

$$t^* = E[t(x)|y, \phi*],$$

and the maximization, or M step, is the determination of the
value of $\phi$, $\phi**$, which is the solution to

$$E[t(x)|\phi] = t^*$$

or equivalently, maximizing the likelihood in which $t(x)$ is
$t^*$.  Since the log-likelihood for exponential families is
convex, the EM algorithm will produce the maximum likelihood
estimate in this case.

Iterative procedures for finding estimators from
incomplete data have been suggested by other authors. Schmee
and Hahn (1979) offer a simple method for regression with
censored data in which each censored data point, $c_x$ (time c

and covariate value x), is replaced by its expected value,
conditional on the observed censored time and previous
estimates of the parameters,

$$E(S|T=c_x) = s_x^* = \int_{c_x}^{\infty} y_x \hat{f}(y_x)dy_x/[1-\hat{F}(c_x)],$$

where $f(y_x)$ and $F(y_x)$ denote the probability density
function and cumulative distribution function respectively,
of the random variable Y at the condition x. The second
stage consists of determining the new estimates for the
parameters by doing a regression on the data set which has
been completed by replacing each censored $c_x$ by its expected
value, $s_x^*$.

The procedure described by Schmee and Hahn determines
the same estimates as are determined by the EM Algorithm.
However, it is easier to describe the Schmee and Hahn method
to non-statisticians and it requires only an ordinary
regression computer package.

An iterative procedure for determining nonparametric
estimates for the conditional survival distribution
$G(S|x) = P(S>s|X=x)$ is given by Finkelstein (1981). In the
models previously presented, X is treated as taking fixed
values, whereas for this procedure, the covariates are
treated as random variables. Thus the likelihood is
factored as

$$L = \prod_i P(X_i,S_i) = \prod_i (X_i=x_i|S_i=s_i)P(S_i=s_i) = \prod_i f_\theta(x_i|s_i)g(s_i).$$

Various models are suggested for the covariate density
$f_\theta(x|s)$, while nonparametric estimates are found for the

survival density g(s). Using an EM Algorithm, this
procedure determines the maximum likelihood estimates for f
and g. This method generalizes easily to left and/or right
censored and tied data.

To illustrate this methodology we consider an example
for which X is a dichotomous covariate. A tractable model
for f when X is dichotomous is the logistic model. For this
model, let $f(x|s)$ be binomial and parametrize
$p_s = P(X=1|S=s)=\exp(a+bs)/[1+\exp(a+bs)]$. For this model,
there is a natural interpretation for the slope parameter b.
The sign of b determines the relative position of the
survival curves for the two groups X=0 and X=1. If b is
negative (positive) then $G_1(s)<G_0(s)$ $(G_1(s)>G_0(s))$ for all
s.

We consider a situation which is commonly encountered
in cancer studies. When studies involve lethal diseases, it
is appropriate to treat the data as right censored and
exact. However, if the disease is nonlethal, and observable
only after death, it is appropriate to treat the data as
left and right censored. The treatment of censored data can
effect the conclusions that are made in an analysis of
failure time data. This is illustrated by the data in Table
1. Lung tumors are nonlethal in RFM mice. It is therefore
appropriate to treat this data as left and right censored.

Although incorrect, we first treat the data as though
the tumors are lethal and calculate the cumulative mortality
curves. The estimates for the parameters for the logistic

model are $\hat{a}$ = 8.59 and $\hat{b}$ = -0.099 and as noted above,
$\hat{G}_0(s) > \hat{G}_1(s)$ for all s.  As can be seen in the graphs shown
in Figure 1, it looks as though the conventional group (X=1)
has a higher incidence or an earlier occurrence of lung
tumors than the germfree group.  If however, we treat the
disease appropriately as nonlethal and observable only after
death, then we obtain the prevalence curves, where G(s) is
the probability that an animal at age s is free of disease.
The estimates of the logistic parameters are now $\hat{a}$ = -1.57

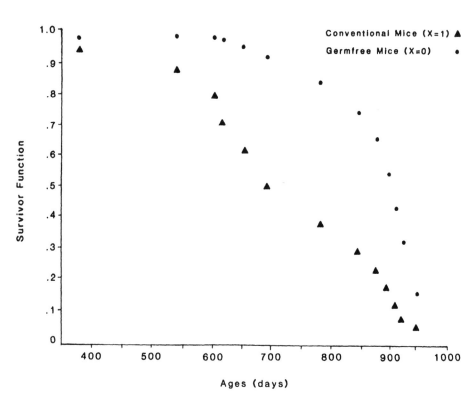

Figure 1    Survival curves (assuming disease length).

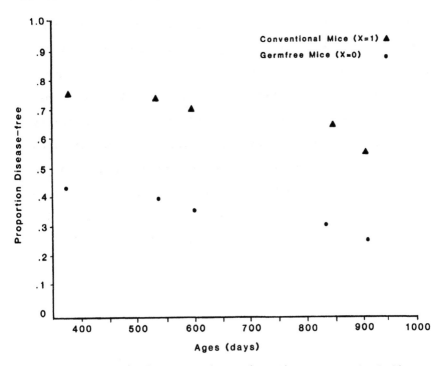

Figure 2    Survival curves (assuming disease non lethal).

and $\hat{b}$ = 0.003 and $\hat{G}_0(s)<\hat{G}_1(s)$ for all s.   As can been seen
in the graphs shown in Figure 2 it is clear that the
incidence is in fact higher in the germfree mice.

Since the treatment of censoring can effect the
analysis of survival data, it is clear that a methodology
for the analysis of doubly censored data is important.

## VI.   CONCLUSIONS

In this chapter, we have presented methods of modeling and
analyzing data which have as an end point the time until an

event occurs. Such data can arise, for example, in studies comparing remission times for patients under two treatments or in followup studies where the time S is age at diagnosis of a lesion. The problem of interest is to determine the dependence of these "failure times" on certain explanatory variables. A secondary interest involves estimation of and specification of models for the underlying failure time distribution. The methods of survival analysis which have been presented here can often provide valuable insight into the information from failure time data which commonly arise from cancer studies.

## REFERENCES

Aalen, A.A. (1978). Nonparametric inference for a family of counting processes. *Annals of Statistics, 6*: 701-726.

Andersen, P.K., Borgan, O., Gill, R.D., and Keiding, N. (1982). Linear nonparametric tests for comparison of counting processes, with application to censored survival data (with discussion). *International Statistical Review 50* (to appear).

Andersen, P.K. and Gill, R.D. (1982). Cox's regression model for counting processes: a large sample study. *The Annals of Statistics, 10*: 1100-1120.

Ayer, M., Brunk, H.D., Ewing, G.M., Reid, W.T., and Silverman, E. (1955). An empirical distribution function for sampling with incomplete information. *Annals of Mathematical Statistics, 26*: 641-647.

Barlow, R.E., Bartholomew, D.J., Bremner, J.M., and Brunk, H.D. (1972). *Statistical Inference Under Order Restrictions*. Wiley, New York.

Breslow, N.E. (1970). A generalized Kruskal-Wallis test for comparing k samples subject to unequal pattern of censorship. *Biometrika*, 57: 579-594.

Breslow, N.E. (1974). Covariance analysis of censored survival data. *Biometrics*, 30: 89-99.

Breslow, N.E. (1975). Analysis of survival data under the proportional hazards model. *International Statistical Review*, 43: 45-58.

Breslow, N. and Crowley, J. (1974). A large sample study of the life table and product limit estimates under random censorship. *Annals of Statistics*, 2: 437-453.

Buckley, J. and James I. (1979). Linear regression with censored data. *Biometrika*, 66: 429-436.

Cox, D.R. (1972). Regression models and life tables (with discussion). *Journal of the Royal Statistical Society* B, 34: 187-220.

Cox, D.R. (1975). Partial likelihood. *Biometrika*, 62: 269-276.

Dempster, A.P., Laird, N.W., and Rubin ,D.B. (1977). Maximum likelihood from incomplete data via the EM Algorithm. *Journal of the Royal Statistical Society* B, 39: 1-22.

Efron, B. (1967). The two sample problem with censored data. *Proceedings of the Fifth Berkeley Symposium in Mathematical Statistics*, IV. Prentice-Hall, New York, pp. 831-853.

Fiegl, P. and Zelen, M. (1965). Estimation of exponential survival probabilities with concomitant information. *Biometrics*, 21: 826-838.

Finkelstein, D. (1981). *A Method of Nonparametric Survival Analysis For Doubly Censored Data*, *Ph.D. Dissertation*, University of Michigan, Ann Arbor.

Gail, M.H. (1975). A review and critique of some models used in competing risk analysis. *Biometrics*, 31: 209-222.

Gehan, E.A. (1965). A generalized Wilcoxin test for comparing arbitrarily singly censored samples. *Biometrika*, 52: 203-223.

Gehan, E.A. (1965b). A generalized two sample Wilcoxin test for doubly censored data. *Biometrika*, 52: 650.

Gehan, E.A. (1969). .Estimating survival functions from the life table. *Journal of Chronic Diseases*, 21: 629-644.

Gill, R.D. (1980). *Censoring and Stochastic Integrals*. Mathematical Centre Tracts 124. Mathematisch Centrum Amsterdam.

Glasser, M. (1967). Exponential survival with covariance. *Journal of the American Statistical Association*, 62: 561-568.

Greenwood, M. (1926). The natural duration of cancer, *Reports on Public Health and Medical Subjects*. (Her Majesty's Stationery Office, London), 33: 1-26.

Gross, A.J. and Clark, V.A. (1975). *Survival Distributions: Reliability Applications in the Biomedical Sciences*. Wiley, New York.

Halley, E. (1693). An estimate of the degrees of mortality of mankind, drawn from curious tables of the births and funerals of the city of Breslau. *Philosophical Transactions of the Royal Society of London*, 17: 596-610.

Hoel, D.G. and Walburg, H.E., Jr. (1972). Statistical analysis of survival experiments. *Journal of the National Cancer Institute*, 49: 361-362.

Kalbfleisch, J.D. and Prentice, R.L. (1973). Marginal likelihoods based on Cox's regression and life model. *Biometrika*, 60: 267-278.

Kalbfleisch, J.D. and Prentice, R.L. (1979). Hazard rate models with covariates. *Biometrics*, 35: 25-39.

Kalbfleisch, J.D. and Prentice, R.L. (1980). *The Statistical Analysis of Failure Time Data*. Wiley, New York.

Kaplan, E.L. and Meier, P. (1958). Nonparametric estimation from incomplete observations. *Journal of the American Statistical Association*, 53: 475-481.

Koul, H., Susarla, V., and Van Rysin, J. (1981). *Multi-Step Estimation of Regression Coefficients in a Linear Model with Censored Data*. Columbia University Technical Report no. B-17.

Lee, E.T. (1980). *Statistical Methods for Survival Data Analysis*. Wadsworth, Belmont, California.

Mantel, N. (1966). Evaluation of survival data and two new rank order statistics arising in its consideration. *Cancer Chemotherapy Reports, 50*: 163-170.

Mantel, N. (1967). Ranking procedures for arbitrarily restricted observations. *Biometrics, 23*: 65-78.

Miller, R.G. (1976). Least square regression with censored data. *Biometrika, 63*, 3: 447-464.

Peto, R. and Peto, J. (1972). Asymptotically efficient rank invariant procedure. *Journal of the Royal Statistical Society*, A, *135*: 185-207.

Peto, R. (1973). Experimental survival curves for interval-censored data. *Applied Statistics, 22*: 86-91.

Prentice, R.L. and Gloeckler, L.A. (1978). Regression analysis of grouped survival data with application to breast cancer data. *Biometrics, 34*: 57-67.

Prentice, R.L. and Marek, P. (1975). A qualitative discrepancy between censored data rank tests. *Biometrics 35*: 861-868.

Schmee, J. and Hahn, G. (1979). A simple method for regression analysis with censored data. *Technometrics, 21*: 417-432.

Tsiatis, A.A. (1981). A large sample study of Cox's regression model. *Annals of Statistics, 9*: 93-108.

Turnbull, B.W. (1974). Nonparametric estimation of a survivorship function with doubly censored data. *Journal of American Statistical Association, 69*: 169-173.

Turnbull, B.W. (1976). The empirical distribution function with arbitrarily grouped, censored, and truncated data. *Journal of the Royal Statistical Society*, B, *38*: 290-295.

Woodbury, M.W., Manton, K.G., and Stallard, E. (1979). Longitudinal analysis of the dynamics and risk of coronary heart disease in the Framingham Study. *Biometrics, 35*: 575-585.

Wu, C.F. (1982). *On the Convergence of EM Algorithm*.
     Technical Report No. 642, University of Wisconsin,
     Madison.

Zippen, C. and Armitage, P. (1966). Use of concomitant
     variables and incomplete survival information in the
     estimation of an exponential survival parameter.
     *Biometrics*, *22*: 665-672.

# 5

# TIME-SPACE CLUSTERING OF DISEASE

George W. Williams
Department of Biostatistics
The Cleveland Clinic
Cleveland, Ohio

## I. INTRODUCTION

The detection of clusters of disease in space, time, or in
both space and time is important to epidemiologist in trying
to explain various health phenomena in a population.  An
aggregation of cases may yield clues as to the causative
mechanism of the disease in question.  Space clustering is a
non-uniform distribution of the cases over the area relative
to the underlying population.  The existence of endemic
pockets of disease can be recognized through a study of
incidence rates for the suspect area.  Time clustering is a
non-uniform distribution of the cases over the duration of
the study.  Pure temporal clustering can be identified by a

study of incidence rates.  Space-time clustering is an
interaction between the places of onset and the times of
onset of the disease, cases which are close in space tending
to be close in time.  Such interaction is regarded as
evidence for contagion or infection of the disease.  The
presence of time-space clustering usually indicates some
causal environmental factor, which is local in both time and
space, such as an infection.  Many of the classic studies
consider only clustering in one dimension, space or time.
For example, the investigation of Snow (1855) of a cholera
outbreak in London might be regarded as a demonstration of
spatial clustering of cases around the Broad Street pump.
However, it is only in recent years that space-time
interaction has been used as a measure of clustering, and,
in fact, has been the subject of considerable research.

   Leukemia has been stated on several occasions to occur
in clusters in space and in time with undue frequency.  If
this could be substantiated, it would be of great
importance, particularly with respect to virus etiology.
Specifically, several clusters of childhood leukemia cases
have come to attention.  A striking example was an
aggregation of eight cases in children under fifteen in
Niles, Illinois, where less than two cases were expected
based on the national average as reported by Heath and
Hasterlik (1963).  The question raises itself whether in a
large country like the United States with tens of thousands
of communities of the size of Niles, Illinois, several

clusters would not be expected to occur at random even if
there were no tendency to cluster. It will be the purpose
of this paper to review the various methods for detecting
clustering in both space and time.


II.  KNOX'S METHOD

A.  Introduction

The question to be answered in a study of time-space
clustering of disease is whether cases which are relatively
close in time are also relatively close in space. Knox
(1963, 1964 a) proposed the study of all possible pairs of
cases and not just successive pairs as had some earlier
authors. Knox's test gave a valid criterion unaffected by
secular variation of diagnosis or notification and
unaffected by the irregular distribution of domiciles.
Playdell (1957) and Pinkel and Nefzger (1959) also suggested
the number of close pairs as an indicator of time-space
clustering.


B.  Method

Knox's procedure begins with a defined geographic area and a
defined period of time, the observed data being the times
and locations of all cases, say n in number, of the disease
under study. Taking all possible $n(n-1)/2$ pairs of cases,
Knox's procedure evaluates whether there is some positive

relationship between the temporal and spatial distance
between the members of a pair. Presumably if there is time-
space clustering, cases in a cluster will be close both in
time and space, while unrelated cases will tend to have a
larger average separation in time and space. Specifically,
pairs of cases can be classified according to both criteria
of proximity in space and proximity in time and a
contingency table constructed as required. The 2x2
condensation is the simplest and perhaps expresses the
nature of the technique most clearly.

If in the complete data $N_{1s}$ denotes the number of
adjacencies (or close pairs of cases) in space and $N_{1t}$ the
number of adjacencies in time, then we may draw up a 2x2
table as in Table 1, where X denotes the number of times
that two points are adjacent in both time and space.

Knox noted that there was a difficulty with this
criterion in the assessment of its statistical significance.
The N observations will not be independent because the same
point will be used many times in determining the

Table 1    Scheme of Knox's 2 x 2 Table

| Time | Space | | |
|---|---|---|---|
| | Adjacent | Not adjacent | Total |
| Adjacent | X | $N_{1t}-X$ | $N_{1t}$ |
| Not adjacent | $N_{1s}-X$ | $N-N_{1s}-N_{1t}+X$ | $N-N_{1t}$ |
| Total | $N_{1s}$ | $N-N_{1s}$ | N |

adjacencies. N, the total number of ways in which the total
number, n, of points may be joined in pairs is equal to
n(n-1)/2. Hence, the standard chi-square analysis of 2x2
tables is not necessarily applicable. Knox (1964 b)
suggested that the number of pairs that are close in time
and close in space can probably be regarded as a Poissonian
variable with expected value calculated in the usual fashion
for that entry in the 2x2 table.

Knox was able to only conjecture the distribution of
his test statistic. David and Barton (1966) showed that his
conjecture was substantially correct and that the
distribution was Poisson in large samples. They gave the
exact mean and variance for Knox's criterion. Using a
graph-theoretic approach, David and Barton (1966) obtained:

$$E(X) = \frac{N_{1s}N_{1t}}{N} = \frac{2N_{1s}N_{1t}}{n(n-1)} \, ,$$

$$E(X^2) = \frac{2N_{1s}N_{1t}}{n^{(2)}} + \frac{4N_{2s}N_{2t}}{n^{(3)}} + \frac{4}{n^{(4)}} [N_{1s}^{(2)} - 2N_{2s}][N_{1t}^{(2)} - 2N_{2t}] ,$$

where

$$n^{(w)} = n(n-1)(n-2) \ldots (n-w+1) \, ,$$

$N_{2t}$ denotes the number of times two time adjacencies
are contiguous,

$N_{2s}$ denotes the number of times two space adjacencies
are contiguous.

The computing of the quantities $N_{1s}$, $N_{1t}$, $N_{2s}$, $N_{2t}$ can be
reduced to simple rote. This is accomplished by preparing a
time graph and space graph. These graphs are prepared by

plotting the points (representing cases) in arbitrary
positions separately in each of these dimensions. Join
those points which are considered adjacent by some defined
criteria. The actual distances drawn do not matter since we
are interested in only the adjacency of points. Such a
diagrammatic representation of a set of points, each pair
either joined or not joined, is technically called a graph.
Consider either the time or space graph and let $\rho_i$ stand for
the number of lines radiating from the $i^{th}$ point
$(i=1,2,\ldots,n)$. Then

$$N_1 = \frac{1}{2} \sum_{i=1}^{n} \rho_i$$

$$N_2 = \frac{1}{2} \sum_{i=1}^{n} \rho_i^2 - N_1.$$

David and Barton (1966) note that to the extent that the
variance equals the mean they confirm Knox's Poisson
contention. Mantel (1967) also gives a general expression
for the exact permutational variance of Knox's number of
close pairs statistic.


C. Adequacy of Approximation

David and Barton (1966) noted that there was confirmation of
the adequacy of the Poisson approximation provided by a
sampling experiment when for Knox's data the time-
coordinates of the cases were randomly allocated to the
space-coordinates and the number of pairs which were
adjacent both in time and space were counted. This process

was repeated many times. The resulting frequency
distribution corresponded quite closely to that of a Poisson
distribution. However, David and Barton's verification of
Knox's procedure does not generally apply, but applies only
to the particular instance with which Knox was concerned, or
perhaps only to similar instances. In other circumstances,
Knox's Poisson procedure could be grossly misleading.


D. Comments

Whether or not there is a positive association between the
two distance measures can, barring important population
shifts during the time period, be assessed without the
population information ordinarily necessary for
epidemiologic investigations. The paired distance approach
has the property of being sensitive primarily to time-space
clustering and being insensitive to clustering which is
purely spatial or purely temporal. This is essential when
notifications are incomplete or erratic, where there is
seasonal variation and where the cases are geographically
clustered in towns. The test does not require knowledge of
the population at risk, its age structure and so on. On the
other hand where such information is available, the test
does not make use of the information and so will not be
fully efficient. This emphasizes the drastic simplicity of
the test, which suggests it will require a large number of
cases to give a reasonable chance of detecting real

interaction, particularly as such interactive effects will
not be very large.

The main difficulty with Knox's method is the
specification of critical times and distances for an
unidentified disease process. The choice of critical
distances in space and time is rather arbitrary, and if the
wrong critical values are chosen, even a fairly strong
tendency to cluster may be missed. On the other hand, if
the critical values are chosen after the data have been
inspected, so as to maximize the apparent significance of
the X statistic, then not only is it invalid to compare X
with the percentage points of a Poisson distribution, but
the distribution of X is unknown and the test becomes
consequently inoperative. To overcome this difficulty Knox
(1963) proposed to divide the time period into intervals and
the range of possible distances likewise and so obtain a
manifold cross-classification. Knox proposed the statistic

$$K = \sum_i \sum_j \frac{(n_{ij} - e_{ij})^2}{e_{ij}}$$

where $n_{ij}$ is the observed number of pairs of cases separated
by a specific time period and by a specified distance and
$e_{ij}$ is the expected value, $e_{ij}$ being calculated in the usual
manner for a chi-square test of independence. He suggested
that it be compared to a chi-square, and, hence, that the
table be regarded as if all entries were independent.

Abe (1973) has shown that Knox's test statistic K
considered as a chi-square variable does not have the

correct expectation. In its place a statistic $Q = \underline{X}'\ V^{-1}\ \underline{X}$ is suggested where the elements of $\underline{X}'$ (a row vector) are the deviations $n_{ij}-e_{ij}$ of the elements of Knox's table from their expectation and $V^{-1}$ is the inverse of the variance-covariance matrix. This latter statistic has the correct expectation. Abe (1973) gives the formula for computing V. An obvious cause of the discrepancy between K and Q is the fact that the pairs of cases considered are not independent.

E.  Pike and Smith's Extension

Knox's method was extended by Pike and Smith (1968) to take into account assumed periods of infectivity and susceptibility of cases. If a disease is contagious with a short latent period, then provided sufficient data are available, the times and places of onset of cases of the disease will display sufficient space-time interaction to be detected by Knox's method. However, if the disease has a long latent interval, Knox's formulation would be unlikely to identify time-space clustering of the disease if, in fact, it existed. Pike and Smith (1968) have extended Knox's method to handle the case of diseases for which a long latent period is postulated and for which associated cases may have their date of onset several years apart. Specifically, for each patient (patient j) with the disease, they postulate two time periods, the first $C_j^p$ during which it is assumed that the patient caught the disease, the period of susceptibility, and the second $I_j^p$ during which the

patient was infective. They also postulate two areas of
space $C_j^a$ and $I_j^a$ representing the patient's effective
movements during his period of susceptibility and period of
infectivity. For a case to be included in the study he must
have been either susceptible and/or infective in the defined
area during the period of the study. Suppose there are n
such cases. These cases fall naturally into one of three
categories, depending on whether they were susceptibles and/
or infectives during the study period and in the study area:

> Category 1: cases who were susceptibles and
> infectives,
>
> Category 2: cases who were infectives only,
>
> Category 3: cases who were susceptibles only.

Suppose there are $n_i$ cases in category i and number the
cases 1 to n in such a way that cases 1 to $n_1$ are in
Category 1, cases $n_1+1$ to $n_1+n_2$ are in Category 2 and cases
$n_1+n_2+1$ to n are in Category 3. Following Barton and
David's (1966) graph-theoretic approach define a directed
space graph, DSG, and a directed time graph, DTG, as
follows. In DSG point k is joined to point j (k→j) if and
only if (iff) $I_k^a$ overlaps $C_j^a$, that is, patient k's area of
infectivity overlaps patient j's area of susceptibility, and
in DTG point k is joined to point j (k→j) iff $I_k^p$ overlaps
$C_j^p$, that is patient k's period of infectivity overlaps
patient j's period of susceptibility. Let X = DSG ∩ DTG,
then X is the number of pairs of patients such that one·
member of the pair was in the right place at the right time

to catch the disease from the other member of the pair. The
larger X the more evidence there is for the given model of
infectivity.

We may display DSG (DTG) as an nxn adjacency matrix A
$(a_{jk})$ $[D(d_{jk})]$ where $a_{jk}$ $(d_{jk})$ is unity iff k→j is DSG [DTG]
and is zero otherwise (by definition $a_{jj}=d_{jj}=0$ for all j),
and X=AD. Moreover, A[D] may be partitioned as

$$A = \left\{ \begin{array}{ccc} A_1 & A_2 & \emptyset \\ \emptyset & \emptyset & \emptyset \\ A_3 & A_4 & \emptyset \end{array} \right\}$$

where $A_1$ is $n_1 \times n_1$, $A_2$ is $n_1 \times n_2$, $A_3$ is $n_3 \times n_1$, $A_4$ is $n_3 \times n_2$ and
$\emptyset$ denotes a null matrix.

Knox's approach may be expressed as a special case of
the above. Let $C_j^a = I_j^a$ be a circle of radius 1/2 δ with
center at patient j's place of onset, $C_j^p$ be from 1/2 τ days
before patient j's date of onset to his date of onset, and
$I_j^p$ be from patients j's date of onset to 1/2 τ days
thereafter, then X is the number of pairs of patients which
satisfy both the distance restriction of having their places
of onset within δ of each other and the time restriction of
having their dates of onset within τ of each other.

The null distribution of X is determined as follows:
the time periods of susceptibility are distributed at random
over the areas of susceptibility and the time periods of
infectivity are distributed at random over the areas of
infectivity with the restriction that the $\{C_j^a, I_j^a\}[\{C_j^p, I_j^p\}]$
are indivisible for j=1,2,...,$n_1$; in other words cases are

randomized rather than the times. With real data the full
evaluation of the randomization distribution of X would be
prohibitively time consuming even with the aid of an
electronic computer. It is necessary either to adopt a
Monte Carlo approach or to approximate the distribution.
Pike and Smith (1968) present a procedure for calculating
$E(X)$ and $\sigma^2(X)$. We may then take $t = [X-E(X)]/\sigma(X)$ as
$N(0,1)$ to approximate the significance level of the observed
X. In general t and the Poisson approximation may be used
as guides to the significance level of X, but a Monte Carlo
approach is needed for complete confidence in the result.

The higher moments of X have been studied by Barton and
David (1966) for the special case of $n_1 = n$, A a symmetric
adjacency matrix, and D an asymmetric (i.e. if $d_{jk} > 0$ then
$d_{kj} = 0$) adjacency matrix, which is the situation obtained
from the original Knox specification. They concluded that
with A and D having a scattered low intensity of 1's, X will
be distributed approximately as a Poisson variate. However,
it is not difficult to construct matrices A and D which give
X very peculiar distributions. Pike and Smith (1968)
discuss the more general situation of $n_1 < n$. The situation
is more complicated than the above but again under
conditions similar to those found by Barton and David
(1966), X will be such that $E(X) \approx \sigma^2(X)$ and we might
conjecture that it will be distributed approximately as a
Poisson variate. The conditions for approximate equality of
mean and variance are roughly that each $A_i [D_i]$ should have a

scattered low intensity of 1's and that $A_i[D_i]$ should be independent of $A_j[D_j]$.

## F. Applications

Knox's test has been utilized in the examination of time-space clustering of several diseases. For example, an important feature of the epidemiology of Burkitt's lymphoma has been the occurrence of "space-time clusters" of patients with the disease in areas endemic for Burkitt's lymphoma. This tendency for patients whose dates of clinical onset of disease are close in time to live closer together than would be expected by chance has given much support to the view that an infective agent may be involved in the etiology of the tumor and that there is a relatively short latent period between infection and clinical onset of Burkitt's lymphoma. [Doll (1978)].

Knox's procedure has also been applied to data on Hodgkin's disease in Connecticut by Kryscio *et al.* (1973) and in Manchester, U.K. by Alderson and Nayak (1971) and results have been negative or inconclusive. Others who have applied Knox's procedure include Lyod and Roberts (1977) and Robert, Lawrence and Lyod (1975).

## III. MANTEL'S GENERALIZED REGRESSION METHOD

Mantel (1967) generalized Knox's procedure for detecting time-space clustering of disease by defining the following statistic

$$Z = \sum_{i<j} \sum X_{ij} Y_{ij}$$

where $X_{ij}$ is a spatial measure between points i and j and $Y_{ij}$ is a temporal measure. Note that Knox's test is a special case where $X_{ij}$, $Y_{ij}$ have measure 1 if cases i and j are within appropriate critical distances of each other. One of many possible distance measures that could be used is to let $X_{ij}$ represent the spatial distance between two cases and to let $Y_{ij}$ represent the temporal distance between two cases. One of the major objections to this measure is that it overly emphasizes the effects of large space and time differences. This objection is based on the fact that if indeed there is a tendency for small time differences and small space differences to occur together, the remaining intermediate and large values of $X_{ij}$ and $Y_{ij}$ are left to align themselves at random. Since the significance of the test statistic depends primarily on whether the large $X_{ij}$ and $Y_{ij}$ are associated, the resultant loss of power is obvious. Mantel (1967) proposed that statistical power can be improved by applying reciprocal transformations to the separations in time and space. Specifically, he let

$$X_{ij} = \frac{1}{k_s + D_{ij}} , \quad Y_{ij} = \frac{1}{k_t + T_{ij}}$$

where $k_s$ and $k_t$ are constants and $D_{ij}$ and $T_{ij}$ are the distance in space and time respectively between cases i and j. The reciprocal transformation gains power over the Knox

procedure since it uses a continuous transformation which
distinguishes between degrees of closeness or distance.


A.  Distribution

The null distribution of Z can be obtained by a finite
population approach analogous to that discussed in the
preceding section.  We have n locations of cases in space
and n locations in time.  The hypothesis of no clustering is
equivalent to one that the locations in space are matched at
random with the locations in time, there being a total of n!
equiprobable sets of matchings.  In principle we can list
the n! possible permutations of our data, compute Z for each
permutation and obtain the null distribution of Z against
which the observed value of Z can be compared.

Were n is too large for the full permutational approach
to be practical, a Monte Carlo approach may be useful.  In
this approach one simulates the randomization scheme enough
times to get the empiric distribution of Z adequate for
significance testing purposes.  For large n an approach can
also be taken which is predicated on the assumption that Z
is approximately normal distributed, so that its deviation
from its null expectation can be tested relative to its null
standard deviation.  Mantel (1967) presents formulas for the
expectation of Z and its permutational variance. While a
full permutational approach can give valid probability
levels for any observed association, for reasons of

practicability it is suggested that the observed association
be tested relative to its permutational variance.

Klauber (1971) and Siemiatycki (1971) found empirically
that sample sizes of the order of a few hundred were not
large enough to guarantee the adequacy of the normal
approximation of Mantel's test in cases where the test
statistic is near the borderline of statistical
significance. For such cases Siemiatycki (1971, 1978) gave
algebraic equations for the derivation of third and fourth
moments. From these, standard Pearson-probability
distributions can be fitted to the data which approximate
more closely the exact distribution than the normal
distribution.

B.  Comments

Siemiatycki (1971) proposed that the best transformation is
that which mimics the probability of a pair of cases being
related.  But since the experimenter does not know a priori
which probability model is operative, an attempt should be
made to give various models a chance to prove themselves.

The use of closeness transformations does not
completely avoid the need for identifying something like
critical distances in space and time.  This is because of
the possibility of zero distances and their corresponding
infinite closeness if the reciprocal transformation is used.
Addition of some constant to the distance before
transforming minimizes the effect of zero or extremely short

distances. However, the results of any analysis could be
affected by the particular constants chosen for addition to
the time and space distances. In the absence of a
reasonable intuitive basis for selecting appropriate
constants, a trial and error approach using a spectrum of
constants may be necessary. However, this upsets the formal
validity of the statistical tests. It is reassuring that
for the reciprocal transformation Glass and Mantel (1969)
showed that over a wide range of arbitrary constants the
test statistic was little changed.


C. Applications

Klauber and Mustacchi (1970) used Mantel's regression
approach to test for space-time clustering for time of
diagnosis and address of 149 leukemia cases in children
under the age of 15 years diagnosed in San Francisco during
the 20 year period 1946 to 1965. The spatial measure chosen
was the distance in feet between the cases which were
plotted to the nearest 100 feet. The temporal measure was
defined as $Y_{ij} = 1$ if cases i and j were diagnosed within
the same time interval and $Y_{ij} = 0$ otherwise. Five
different time intervals were investigated:  0.5, 1, 2, 9,
and 12 months. In order to avoid the difficulties of
repeated testing, the 2-month interval was the predetermined
value chosen for significance testing purposes. The census
data for San Francisco indicated that the rate of population
growth varied considerably from census tract to census

tract.  In order to reduce the effect of this artificial
clustering in the population at risk, the case series was
analyzed in separate five year periods.  For the
predetermined time interval of 2 months for ages 0 to 14
years a nonsignificant p-value of 0.32 was obtained.
Additional analyses with reciprocal spatial and temporal
measures were performed.


IV.  PIKE AND SMITH CASE CONTROL METHOD

Pike and Smith (1974) suggest an approach for investigating
time-space clustering of disease using information derived
from a set of matched control persons as well as the cases
themselves.  The case-control approach attempts to ascertain
whether patients with the disease have had more relevant
contact with each other than have a suitably chosen control
group.


A.  Method

Let $X_{ij}$ = 1 if effective contact between cases i and j,
        = 0 otherwise.
Consider the statistic

$$Z = \sum_i \sum_j X_{ij}$$

where the summation is over all possible pairs of cases.

      If a matched control subject is chosen for each
patient, there will be n sets of persons to be interviewed,
each set consisting of one patient and his control, or 2n

persons in all. We assume that the date of onset and
clinical course of hypothetical disease for a control is the
same as that of his patient's disease; and that contact may
be measured, under the postulated rules of effective
contact, between all possible pairs of persons. The null
distribution of Z may then be obtained by selecting n
persons from the total of 2n persons by taking one person at
random from each set (either a case or a control) and
measuring the total contact, Z, between these n persons.
There are $2^n$ possible different random selections with
associated values of Z and the statistical significance
level of the observed value of Z may in principle be
evaluated exactly.

We may generalize the problem by considering each set
not to involve only one patient and one control but to
consist of a number of patients and controls. That is, the
total of n patients and m controls are divided into k
strata, the $i^{th}$ stratum consisting of $n_i$ patients and $m_i$
controls where

$$\sum_1^k n_i = n; \quad \sum_1^k m_i = m; \quad t_i = n_i + m_i; \quad \sum_1^k t_i = t.$$

This would arise, for example, if the control sample of size
m were drawn without reference to which cases they matched
and the cases and controls were then stratified by age, sex,
etc. Allowance could then be made for a group of such
factors in the analysis of the data.

For this general situation let $X_{ivjw}$ be the measure of
contact from the $v^{th}$ patient in stratum i to the $w^{th}$ patient

in stratum j. The sum of these measures amongest all
possible pairs of the n patients is

$$Z = \sum_{i=1}^{k} \sum_{v=1}^{n_i} \sum_{j=1}^{k} \sum_{w=1}^{n_j} X_{ivjw}$$

where $x_{iviv} = 0$ by definition.  The null distribution of Z
is obtained by selecting n persons at random from the total
of t persons with the restriction that $n_i$ persons are taken
from the $t_i$ persons in the $i^{th}$ stratum.

An approximate significance level may be obtained by
simulation, or by approximating the distribution of Z by
evaluating its expected value, E(Z), and variance, V(Z), and
considering

$$\frac{[Z-E(Z)]^2}{V(Z)}$$

as chi-square with one degree of freedom or equivalently
considering

$$\frac{Z-E(Z)}{\sqrt{V(Z)}}$$

as a normal random variable.  Pike and Smith (1974) provide
formula for E(Z) and V(Z).

In both the generalized Knox approach and the case-
control method, the assumption of a chi-square distribution
to calculate the level of significance should be a fair
guide to the true level.  However, it is not difficult to
construct data sets which may give misleading results.

The case where there is only a single stratum has been
examined by Mantel and Valand (1970) and Mantel and Bailar
(1970).  Pike and Smith's procedure generalizes this earlier

work to cope with heterogeneity in the study population. It
is because of such heterogeneity that the null distribution
of Z is not given by a random selection of n person out of t
persons, but rather by a random selection of $n_i$ out of each
$t_i$. It should be noted that Mantel and Valand (1970)
incorrectly claimed that the sampling distributions of their
statistics were approximately normal and, as a consequence,
their inferential procedures are incorrect. Mielke (1978)
and Mielke *et al*. (1976) describe the limit theorem
condition for asymptotic normality which is not satisfied by
the Mantel and Valand statistics and provide approximate
inferential results for the example given by Mantel and
Valand.


B.  Comments

The best epidemiological evidence of person-to-person
transmission of a disease is the occurrence of direct
contact of one person with another at a relevant time. By
careful questioning, one could specify where each patient
had been during their period of susceptibility and
infectivity. However, an overlap of one patient's area of
susceptibility with another's area of infectivity and also
an overlap of their respective susceptible and infective
time periods does not necessarily imply that they were ever
in the same place at the same time. This aspect should be
considered when applying Pike and Smith's method.

Secondly, relevant contact is dependent upon postulating susceptible and infective periods for each patient. The specifications of these periods are also critical. Pike and Smith's approach is likely to be of most value when the infective and susceptible periods may be well defined, as would be the case if the susceptible period was in utero and patients were infective around the date of onset of the disease.

Finally, as is true for all case-control studies, the selection of a satisfactory control group is crucial.

Pike and Smith's procedure has been applied in a study of clustering of cases of leukemia, Hodgkin's disease, and other lymphoma's in Bahrain by Hamadeh *et al.* (1980). No significant clustering was found.

V.   TWO SAMPLE PROBLEM

Mantel's generalized regression approach has been extended to the two sample case by Klauber (1971). In the two-sample situation one is confronted with two sets A and B of distinguishable points (or cases) having space coordinates and time coordinates. One may wish to detect whether or not there is a tendency for pairs of points, one from A and one from B, which are spatially close to be also temporally close and conversely. Such a case might be an epidemiologic problem where one suspects a common origin of a disease in two species, and there exists surveillance for the disease in both populations.

As has been true for other time-space clustering
methods discussed, the methods described below will not
depend on knowledge of the underlying spatial or temporal
distributions of the populations A and B from which the
sample sets of points may have been identified.


A.  Method

Let $(X_m, Y_m, T_m)$, $m=1,\ldots,n_1$, and $(X_n^*, Y_n^*, T_n^*)$, $n=1,\ldots,n_2$,
represent the space and time coordinates of two complete
sets of point observations A and B, respectively, in a given
region and time interval.  Let $S_{mn}$ and $R_{mn}$ be real-valued
functions of the space and time distances, $D_{smn}$ and $D_{tmn}$
respectively, between case m from set A and case n from set
B, i.e.,

$$D_{smn} = [(X_m - X_n^*)^2 + (Y_m - Y_n^*)^2]^{1/2} ,$$
$$D_{tmn} = |T_m - T_n^*| .$$

The sign of the time differences may have to be taken into
account in the function R if one set of events is considered
a possible antecedent to the other.  Frequently used
functions are the reciprocal distance measures
$S_{mn}=(D_{smn} + b_s)^{-1}$ and $R_{mn}=(D_{tmn} + b_t)^{-1}$ where $D_{smn}$ and $D_{tmn}$
represent the space and time distances, respectively,
between cases m and n and $b_s$ and $b_t$ are specified constants.
Also the indicator function corresponding to Knox's test in
the one sample case is used with

$S_{mn}$ = 1 if $D_s < c_s$ ,

     = 0 otherwise ,

$R_{mn}$ = 1 if $D_t < c_t$ ,

     = 0  otherwise.

The statistic $Z = \sum_m \sum_n S_{mn} R_{mn}$ may be compared to its randomization distribution by assuming that one of the sets is fixed and the coordinates of the other are randomly permuted or that both sets are random. An observed value of Z may be compared to its appropriate distribution (a) assuming the A time coordinates are randomly permuted, the B time coordinates and both sets of space coordinates are fixed, or (b) assuming that both sets of space coordinates are fixed and within each set, the time coordinates are permuted at random.  Theoretically, it would be possible to compute all values of $D = Z - E(Z)$ and by enumeration compute the probability under the null hypothesis that D is as great or greater than the observed value and similarly for the negative and two-sided alternatives.  However, for only small values of $n_1$ and $n_2$ will it be feasible to compare D to its complete distribution under either randomization distribution.  Randomization experiments using a computer may be used to compare Z with its empirical distribution.

In some instances, the approximate standardized normal deviate $u = [Z - E(Z)]/\sigma_Z$ may be used.  If the random points are independent and the function $Z_{mn} = S_{mn} R_{mn}$ has finite second moment, then the central limit theorem for sums of

this type implies the asymptotic normality in the case of
only one random set. If, in addition, both sets of time
coordinates are random, asymptotic normality obtains by
Lehmann's (1951) extension of Hoeffding's (1948) U statistic
theory. Klauber (1971) obtains the mean and variance of Z
for both randomization models.

B. Power and Approximations

Klauber (1971) investigated the adequacy of the
approximation and concluded that in some situations one
should be able to use the normal approximation. However, it
is recommended that a randomization experiment be performed
in any new situation and whenever normal critical levels are
borderline say within $\pm$ 0.03 of the level of significance.

The proposed test has power to detect clustering using
moderate sample sizes. In fact, one should be cautioned
regarding the high degree of sensitivity of the test. A
high level of statistical significance is possible even when
few points are involved in the clusters.

C. Application

As noted by Klauber (1971), this procedure could be applied
to data representing the location and time of diagnosis of
lymphatic leukemia cases in both man and pet cats. This
method could be used to gather evidence of contagion (or
lack of contagion) from cat to man, man to cat, or of both

types.  Pet cats are of particular interest since a virus
particle has been implicated as the agent in feline
leukemia.


## VI.  SEVERAL SAMPLES

The two-sample randomization tests for space-time clustering
of Klauber (1971) have been extended to cases with more than
two samples by Klauber (1975).  These tests are sensitive to
a tendency for pairs of points (no two members of each pair
from the same set) which are close in space to be close in
time and conversely.  Such tests are insensitive to
clustering of the points in space or time alone or space-
time clustering within each set.  That is, a between sets
space-time clustering test is developed.  As in the two-
sample case, the basic approach is similar to that of
Mantel's (1967) one-sample test.


## A.  Method

Assume q sets of space and time locations $\underline{s}_1,\ldots,\underline{s}_q$ of sizes
$n_1,\ldots,n_q$ are obtained.  Space and time measures between two
points, the $m^{th}$ point in $\underline{s}_i$ and the $n^{th}$ point in $\underline{s}_j$ are
denoted by $S_{mn}^{ij}$ and $R_{mn}^{ij}$ respectively.  Superscripts to
the functions will be used to indicate pairs of sets and
subscripts will be used to indicate pairs of points within
the corresponding sets.  As in the two-sample case of
Klauber (1971), the measure of space-time interaction

between $\underline{s}_i$ and $\underline{s}_j$ is given by

$$Z^{ij} = \Sigma\Sigma \underset{mn}{S^{ij}} \underset{mn}{R^{ij}} \; , \quad m=1,\ldots,n_i \; , \quad n=1,\ldots,n_j \; .$$

The overall measure may be computed as

$$Z = \underset{i<j}{\Sigma\Sigma} Z^{ij}$$

where the summation may be taken over all i and j, i < j, or only for those i and j, i < j, for which at least one of $\underline{s}_i$ or $\underline{s}_j$ is random. The q-sample test provides the opportunity to test up to $2^q - 1$ different models, depending on whether one or more of the q samples are considered random.

The means and variances of the $Z^{ij}$'s may be obtained from Klauber (1971). The remaining expressions for $E(Z^{ik}Z^{jk})$ are obtained in Klauber (1975). A normal approximation for Z can be used. The one-sample approach of Mantel (1967) provides the within sets and grand total analysis. The two-sample approach of Klauber (1971) yields the three 2-sample "between sets" analysis and the above method provides the total between analysis. The analyses can be summarized in an "analysis of clustering" table comparable to tables for the analysis of variance.

## B. Application

An example could be a situation where the evidence of leukemia cases are recorded over time in a region in different animal species. In particular, one might be interested in testing a model where feline cases would be

considered fixed and cases in other species, random, since a
leukemogenic virus has been identified in cats. An example
using actual human, canine, and feline leukemia and lymphoma
is provided by Klauber (1975).

VII.  RIDITS (ONE SAMPLE AND TWO SAMPLE)

A somewhat different approach to the problem of time-space
clustering of disease has been presented by Pinkel *et*
*al*. (1963).  The approach by Pinkel *et al*. (1963) is based
on the assumption that cases more than some critical
distance apart are probably not related.  The distribution
of time distances between pairs where the spatial distance
exceeds the critical distance provides a reference
distribution against which time distances for closer pairs
are grouped.  Distances for each close pair are categorized
and the temporal distribution for each category is
summarized as an average ridit.  A ridit is simply the
percentile standing of individuals in a group under study
relative to an identified distribution or reference
distribution.  In fact, the first three letters of ridit
stand for relative to an identified distribution.  Bross
(1958) describes in some detail how to use ridit analysis.
Confidence intervals for the average ridit in each of the
categories are computed on the basis of the observed
variance in ridit values for pairs falling within the
distance category.  No specific statistical significance
test is given, but an indication of clustering can be gained

from an examination of average ridits and their confidence
intervals for the distance categories. The ridit approach
has been recently employed in a study of multiple sclerosis
by Poskanzer *et al.* (1981).

This approach can be extended to the two-sample case.
In the two-sample case, cases taken from each sample more
than some specified distance apart are assumed to be
unrelated and the approximate empirical cumulative
distribution function of the corresponding time distances is
considered the relatively identified distribution. The
following two-sample space-time clustering tests were
performed by Pinkel *et al.* (1963) - leukemia cases versus
solid tumor cases, leukemia cases versus traffic fatalities,
and solid tumor cases versus traffic fatalities. Traffic
fatalities were analyzed in order to show the absence of
clustering in situations where it would not be expected.
The authors interpreted their results as indicating
clustering of leukemia deaths and of solid tumor deaths, and
also association of these two causes of death for spatial
distributions under 1/8 of a mile. Clustering is not
indicated for traffic fatalities.


## VIII.  EDERER'S METHOD

As opposed to the procedures for detecting time-space
clustering of disease discussed above, the procedure of
Ederer *et al.* (1964) does not make explicit use of the
paired distance technique. This procedure examines the

distribution of cases within a time-space unit. For
purposes of illustration and in keeping with the paper by
Ederer *et al.* (1964), let us consider a time unit of 5 years
and a spatial unit of a town. With a single case there can
be no cluster, hence we confine the study to time-space
units with two or more cases. For units with a total of
exactly two cases, the two cases can distribute themselves
over the five years in two distinct ways: (a) they can both
occur in the same year, or (b) they can occur in a different
year. In terms of ordered occupancy numbers, we can have
two types of distributions: (a) 2, 0, 0, 0, 0 or (b) 1, 1,
0, 0, 0. If an unusually large proportion of 5-year-town
units with two cases display distribution (a), this would be
an indication that the cases have a tendency to cluster.


A. Method

As an index of clustering we consider the statistic $m_1$, the
largest of the occupancy numbers [e.g. $(r_1, \ldots, r_n)$] where $r_i$
is the $i^{th}$ occupancy number when the cells are arranged in
descending order of the occupancy number
$r_1 \geq r_2 \geq \ldots \geq r_n$. We can easily obtain the conditional
distribution of $m_1$ given

$$r = \sum_{i=1}^{n} r_i$$

based on the work of Feller (1957). He presents a method
for determining the probability distribution for r balls in

n cells.  For example, for n = 5, the probability of a set

of occupancy numbers is

$$P(r_1, r_2, r_3, r_4, r_5) = \frac{r!}{r_1!r_2!\ldots r_5!} \ \frac{5!}{n_0!n_1!\ldots n_5!} \ (\frac{1}{5})^r$$

where $n_j$ is the number of cells containing exactly j balls.

Here,

$$\sum_{j=0}^{r} n_j = 5 .$$

From the conditional distribution of $m_1$ given r, one can

obtain the means and variances of the $m_1$ values.  Then

normality can be assumed for $\Sigma m_1$, the sum of all empirical

annual clusters across all 5 year periods and across all the

towns.  Then a continuity-corrected chi-square with one

degree of freedom is computed:

$$\chi_1^2 = \frac{(|\Sigma m_1 - E(\Sigma m_1)| - 0.5)^2}{\Sigma V(m_1)}$$

Ederer *et al.* (1964) present expectations and variances

of $m_1$ for totals up to 15 cases distributed among three,

four, or five time periods.  Mantel *et al.* (1976) present

tabulations of expectations and variance for up to 500 cases

distributed among two or three time periods and 200 cases

distributed among four or five time periods.  Asymptotic

formulas are provided for the expectation and variance of

the largest frequency in a single period when arbitrarily

many cases are distributed at random among two, three, four,

or five time periods.

B.   Related Results

This method can be extended to consider $m_2$, the largest
total in two successive years, or more generally to $m_t$, the
largest total in t successive years.  As r gets large,
enumerating the occupancy distributions become very time
consuming.  For large values of r and t the functions $E(m_t)$
and $V(m_t)$ could be estimated from occupancy distributions
generated by Monte Carlo methods.  Mantel et al. (1976) have
suggested several other functions of the occupancy numbers
that could be used for testing for time-space clustering of
disease.

In considering temporal clusters the calendar year was
chosen as the unit of time for the example above, and in
this way obtained 5 non-overlapping years in a 5 year
period.  The reader will recognize that a 5 year period in
fact contains a continuum of overlapping periods one year in
length. Ederer et al. (1964) note that some work has been
done on computing the probability of clusters in such a
continuum.  In fact, Naus (1966) and Wallenstein (1980) have
discussed the scan statistic which is the maximum number of
observed cases in an interval of preselected length, as the
interval is allowed to scan, or slide along, the time frame
of interest.  Wallenstein (1980) presents tables which will
allow for easy use of the statistic.

C.   Power

To investigate the power of the method, Ederer et al. (1964)
considered whether the method was sufficiently sensitive to

detect clustering in some disease of known viral origin.
The evaluation indicated that the method was quite powerful
for detecting clusters in samples when the clustering is as
intense as that occurring for polio or hepatitis.

D.  Comments

Ederer *et al.*'s method is sensitive both to temporal and to
temporal-spatial clustering.  Data would have to be examined
to see which is occurring.

E.  Application

Ederer *et al.*'s procedure has been applied to data from
Connecticut [Ederer *et al.* (1965) and Kryscio *et
al.* (1973)], Michigan [Stark and Mantel (1967a, 1967b)], and
Manchester, U.K. [Alderson *et al.* (1971)] and results have
not indicated clustering.

IX.  DAVID AND BARTON'S METHOD

David and Barton (1966) and Barton *et al.* (1965) present
another test for time-space clustering.  A point of
dissatisfaction which David and Barton raise with respect to
the Knox approach is its dependence on the critical distance
in time and space that is selected.  Also, information
contained in the actual distances between two points is lost
by the Knox approach.  To remedy this they suggest applying
a test they had previously derived for studying the
randomness of points on a plane.

Their test is developed in analogy with the analysis of
variance F test. Suppose there are n points in a plane,
between any two of which we can measure a distance, getting
the total of squared distances between the n(n-1)/2 possible
pairs. If the n points can be divided into subgroups,
similar sums of squares can be obtained within each
subgroup. By subtraction, a residual for "between
subgroups" is determined and a test is based on the between-
to-within subgroups ratio. The test is applied to the
disease incidence data by dividing the study period into
subintervals each subinterval identifying a subgroup for
this test, and analyzing the distribution of the points in
space. Specific objective rules for identifying the
subintervals are given. A statistical weakness of the
David-Barton test is related to its multi-degree of freedom
character.

Mantel (1967) notes that the David-Barton (1966)
approach could readily have been turned around so as to
correspond to the standard analysis of variance technique.
Consider the location map of all disease cases. By
inspection of the map, and without reference to the times of
occurrence of each case, divide the area into geographic
subregions, with cases in a subregion thus close in space.
An analysis of variance can now be performed on the times of
occurrence, with variation subdivided into "between" and
"within" subregions. Time-space clustering is indicated by
a significantly large "between" component.

X.  GENERAL COMMENTS

Whenever there is significant evidence of clustering, one
must look into the possibility of artifactual causes such as
differential population shifts, changes in medical practice,
etc.  Significant time-space clustering does not necessarily
imply that a disease was transmitted from one case to
another, only that two cases may sometimes have a common
source.  This could be interpreted as due to a toxic rather
than an infective agent, spread through atmospheric
pollution, or contamination of food or water supplies; or as
a result of direct contact with toxic weed killer, paints,
solvents, and other poisonous materials.  If clustering is
detected, this may be due to reasons other than contagion as
for example where there has been more extensive reporting in
one part of the region under investigation during the early
part of the period of study whereas later in the period
another region has been more fully reported.  Knox (1964)
pointed out that a high time-space concentration could be
the result of a movement of space clusters in time.  If
there were a population migration during the period of
study, there is a possibility that this would lead to a
significant result and spurious clustering would be
detected.  To counter this possibility, one could subject a
control series to the same analysis as the cases.  Any
migratory trends would result in a high number of pairs of
controls close in both time and space.  The controls should
be matched for those variables that might influence the

tendency of people to move.  Heath (1977) performed just
such an analysis in his study of cleft lip and palate in the
Oxford area.  Alternatively, in order to minimize the effect
of trends and shifts in the population, it is possible to
divide the data into separate periods of time and perform
the analysis for time-space clustering separately for each
period.

The absence of clustering has certain implications for
an infectious hypothesis.  It is not compatible with any
close time or space relationship due to person-to-person
transfer of the agent and its effects.  On the other hand,
it would not exclude the action of some virus or other
infective agent latent in the environment.  On the
application of the methodology for detecting disease
clustering a caution in its use becomes apparent.  Suppose
that in some region there is an inordiante number of cases
of disease, unquestionably a cluster.  We now take several
years' data for the region and apply the method.  It is not
unlikely that no significant clustering will be found.  The
explanation is that we know we were dealing with a time-
space cluster because the same kind of thing was not
happening elsewhere, but we have not fed this information
into our analysis.  To do so would have required obtaining
data not for the region in question only, but rather for a
larger area of which the region is only a part.  Without the
additional information the data would have been consistent
with simple temporal clustering.

The power of tests for time-space clustering of disease
is affected by the definition of the endpoint. If the
possible infectious nature of a disease is the subject of
study, then ideal data in the time dimension would give the
instants of infection, but in practice the best that can be
obtained are data on the times of diagnosis, in turn
occurring some time after the times of symptom onset; if the
latent period or the diagnosis time are highly variable,
then a test for space-time interaction may be subject to a
considerable loss of power. In other studies some outcome
of the disease such as death may provide the time data, and
if the outcome is probabilistic in nature, again tests for
clustering of cases may be weakened. For example, two
related cases of leukemia could have time of genesis close.
With time of diagnosis the endpoint, related cases could
differ somewhat. Related cases of leukemia could differ
considerably in times of death. Related cases could succumb
at widely disparate times. Often the information on
infective times is rather poor; this is so in many diseases
in man with a somewhat variable latent period, in community
health surveys where the retrospective nature of the data
makes it unreliable, or in surveys of animal populations
where there are no case histories at all.

It is of interest to note that in 1971, Vianna and his
co-workers reported an aggregation of Hodgkin's disease
patients centered about a particular high school in Albany,
New York. The Albany study is of great importance in that

it used methods of infectious disease epidemiology to search
for aggregation of exposures rather than clustering at the
time of disease manifestation.  It also demonstrated that
clusters might be defined in terms of both direct and
indirect contact with cases.  The failure to find strong
evidence of clustering does not, however, weigh heavily
against positive evidence for the existence of infective
agents obtained in the laboratory or the ward.  Statistical
techniques to allow for different durations of infectivity
and different lengths of incubation time are clumsy in the
absence of any clinical indications of what the periods are
likely to be.  More importantly, the possibility that
clinical infection may occur without the appearance of overt
disease makes the techniques inefficient, so that they may
fail to detect evidence of clustering of diseases that are
known to be infectious, such as infectious  mononucleosis.

Experience leads us to believe that for a disease of
uncertain etiology, Mantel's method should be used with a
number of transformations of which Knox's method is one.
Various critical time and distances could be used although
the a priori specification of a critical time and distance
for significance testing is advised.  The normal
approximation may be used, but in the case of near
significant results an attempt should be made to estimate
the exact distribution.

The findings based on these analyses in the cancer
literature are as follows.  There have been few studies of

the leukemias and the lymphomas as a group and most of the
studies of case-clustering have been conducted on leukemia,
Hodgkin's disease or Burkitt's lymphoma. Studies of space-
time clustering involving patients with leukemia and
Hodgkin's disease have not yielded convincingly positive
findings. However, one might not expect to find such
clustering of cases if the period between "infection" and
onset is long. A special study of childhood leukemia [Smith
*et al*. (1976)], assuming a possibly long latent period did
not produce evidence of case-to-case transmission. The
studies of Hodgkin's disease are conflicting. Strong
evidence of clustering of cases in schools has been reported
in one U.S. study [Vianna *et al*. (1971, 1972)] but a similar
British study [Smith *et al*. (1977)] did not find this. The
evidence of Vianna and his colleagues (1972) of contact
between 31 people with Hodgkin's disease in Albany,
sometimes directly and sometimes indirectly through an
intermediary, is difficult to evaluate. In principle, it
can be assessed only by using objective criteria to define a
contact and then comparing the contacts with those in a
control series that is investigated equally intensively.
When Smith *et al*. (1977) used this method in Oxford, they
failed to find any greater number of contacts between
patients developing Hodgkin's disease under 40 years of age
than between a random sample of patients of the same age and
same social class, treated in the same hospitals during the
same period. Further investigations are necessary to

determine whether case-to-case transmission may take place.
There is indirect evidence which suggests that Hodgkin's
disease might arise as a rare response to a common infection
[Glass and Mantel (1969)]. In such circumstances case
clustering of Hodgkin's disease would not necessarily be
expected to occur. Space-time clustering of Burkitt's
lymphoma has been a distinctive feature of the epidemiology
of this disease though it has not been seen in all areas
where Burkitt's lymphoma is endemic. [Morrow *et al*. (1970),
Brubaker *et al*. (1973), and Morrow *et al*. (1977)]. In a
recent study [Williams *et al*. (1978)], support was given to
the notion that the clustering is a real effect and does not
represent an artifact of case ascertainment. More recent
studies of this disease include Semiatycki *et al*. (1980).
For additional details regarding the assessment of "case
clustering" of lymphomas and leukemias, see Smith (1978).


XI.  APPLICATIONS TO THE PROBLEM OF DETECTING HOUSEHOLD
     AGGREGATION OF DISEASE

A.  Introduction

We shall now consider a particular type of spatial
clustering, that within households. It is clear that
household aggregation of disease may arise through either
environmental cr genetic influences and in general it may be
difficult or impossible to separate out these components.
Nevertheless, the detection of such clustering may be of
considerable value in studying disease etiology.

Most of the common cancers show some degree of familial clustering. However, for the common cancers the risk among siblings of cancer patients is seldom more than twice that in the general population, and we cannot exclude the possibility that an excess of this degree is due to a common environment. Even childhood cancers, which might be expected to have a large genetic component show only a minimal tendency for familial clustering. For many diseases, including cancer of specific primary sites, a variety of associations with environmental factors and personal attributes of individuals have been demonstrated. Moreover, the study of the degree of concentration of cases in households will be important in assessing the importance of household or family common factors as well as family contact in the causation of disease, even though it does not distinguish the latter from the former factors.

Statistical techniques to help decide whether small clusters of rare diseases are associated or merely reflect chance occurrence have been greatly improved since Knox's method was first suggested (see Section II). We will review the application of the various methods for detecting time-space clustering of disease to the problem of testing for familial aggregation, starting with the method proposed by Mathen and Chakraborty (1950).

B. Mathen and Chakraborty's Method

Mathen and Chakraborty (1950) have investigated the probability distribution of the number of households

affected if disease occurs at random.  If $m_1$, $m_2$,... are the
sizes of the households, n the total number of cases
observed in the population studied, and s the total number
of households affected, then the probability distribution of
s will serve as a basis of assessing the importance of the
household aggregation of disease.

First consider the case when the size of the households
is constant and equal to m.  Mathen and Chakraborty (1950)
derive the distribution of s.  However, the calculation of
the levels of significance of s in the case of equal-sized
households is laborious.  Therefore, an approximate test of
significance in the large sample case has been developed.

Consider r households with

$$\frac{n}{mr} = 1 - k = \text{proportion of people affected by disease.}$$

As r tends to infinity keeping n/mr constant, the
distribution of s tends to the normal form.  The convergence
to normality is slower for values of k in the neighborhood
of 1 or 0, i.e., when the proportion of people affected is
very small or very large.

Even when the family sizes are different, the
distribution of s tends to the normal form with increasing
size of sample.  Suppose that in the observed sample there
are $r_i$ families of size $m_i$, that $n_i$ cases occur in $s_i$
families, i = 1,2,...,u, and that the proportion of affected
people is neither too small nor too large.  Consider the n
cases which are distributed randomly among the r families of

the sample to be divided into sets of $(n_1, n_2, \ldots, n_u)$ cases
such that for any particular set each $n_i$ is distributed
randomly among the group of $r_i$ families of size $m_i$. When
the numbers $r_i$ and $n_i$ are fixed at the values actually
obtaining in the sample, $s_i$ may have different values as the
$n_i$ cases may be distributed among $r_i$ families in different
ways. For family group size $m_i$, we have to consider the
distribution of $s_i$ for given $r_i$ and $m_i$. We have already
seen that the distribution of $s_i$ tends to the normal form
for large $r_i$, $n_i/m_i r$ remaining a constant. The total number
s of affected families in the sample is given by
$s = s_1 + s_2 + \ldots + s_u$. Provided the total number of families
investigated is large enough to make $r_i$ sufficiently large,
each $s_i$ will be approximately normally distributed. The
actual expressions for the conditional expectation and
variance of s if the sample consists of $r_i$ families of size
$m_i$ where $r_i$ cases occur in $s_i$ families, $i = 1, \ldots, u$ are
given by Mathen and Chakraborty (1950). Therefore,
knowledge of the conditional expectation and variance of s
in the case of unequal sized families gives us a method of
ascertaining in large samples whether family common factors
or family contact predominate in the causation of disease.
Thus, if in a certain case it is found that the actual
number of families affected is less than the expected, it
may be inferred that family contact or family common factors
are significant in the causation of disease.

In summary, Mathen and Chakraborty (1950) consider the
distribution of disease within a community of households,

and regarding the total number of cases as fixed, use as a
test statistic the number of households containing at least
one case. Evidence of the infectious nature of a disease is
provided by a significant deviation of the statistic from
its null expectation in the direction of small values, but
the test may not be very powerful because of the loss of
information on the exact number of cases within each
household.

## C. Walter's Test

As noted earlier the test developed by Mathen and
Chakraborty (1950) does not use information on the exact
number of cases within each household. The test developed
by Walter (1974) uses this information and is shown to be
rather more powerful as a result. In addition, the
calculation of moments and the treatment of single
ascertainment data are somewhat simpler than in the Mathen
and Chakraborty test.

Suppose that a population of n individuals is grouped
in s houses containing $m_1, m_2, \ldots, m_s$ members: thus $\Sigma\ m_i = n$.
Further we suppose that the $i^{th}$ house contains $r_i$ cases of
infection, $0 \leq r_i \leq m_i$, with a total of $r = \Sigma\ r_i$ cases
regarded as fixed. Considering now the $n(n-1)/2$ possible
pairs of distinct members of the population, we take as out
test statistic those pairs coming from the same household,
both being infected, and denoted by $N_r$.

|  | Both individuals infected | Not both individuals infected | Total |
|---|---|---|---|
| Both individuals from the same household | $N_r$ | | |
| Individuals from different households | | | |
| Total | | | $n(n-1)/2$ |

Note that Walter's test is a special formulation of the test proposed by Knox (1964 a (see Section II)). To express Walter's problem in the same form as that of Knox, let all r cases have a pseudo onset time $T_0$. Let the remaining (n-r) unaffected persons have (n-r) distinct pseudo onset times at $T_1, T_2, \ldots, T_{n-r}$ all different form $T_0$ and from each other. Then $N_r$ is identical to Knox's X with critical time $\tau = 0$ and critical distance $\delta = 0$. The formulation of the problem allows one to use the formulas of Mantel (1967) for the mean and variance of $N_r$ as mentioned in Section II B. The higher moments and asymptotic distribution of $N_r$ and the power of the test have been considered by David and Barton (1966) and Barton *et al*. (1965) and were presented in Section II B. In the notation of this section,

$$N_r = \sum_{i=1}^{S} r_i(r_i-1)/2 \ ,$$

$$E(N_r) = \frac{r(r-1)}{2n(n-1)} \{S_2-n\} \tag{1}$$

where

$$S = \sum_{i=1}^{s} m_i, \quad S_2 = \sum_{i=1}^{s} m_i^2, \quad S_3 = \sum_{i=1}^{s} m_i^3,$$

$$
\begin{aligned}
Var(N_r) = r^{[2]} \ (n-r) \ [2n(n-1) \ (r-2)S_3 \\
+ \ S_2\{n(n+1) \ (n-r-1) + S_2(3n+3r-3-2nr)\} \\
- \ n^3(n-r-1)/2n^{[4]} \ n^{[2]}
\end{aligned}
\tag{2}
$$

and

$$n^{[v]} = n(n-1) \ (n-2)...(n-v+1) \ .$$

If the community consists of s houses with $m_i = m$, then

$$E(N_r) = \frac{r(r-1) \ (m-1)}{2(n-1)} \ ,$$

$$Var(N_r) = \frac{r(r-1) \ (n-r) \ (n-r-1) \ (n-m) \ (m-1)}{2(n-1)^2 \ (n-2) \ (n-3)} \ .$$

It may be noted that the calculation of these moments is considerably simpler than the corresponding statistics for the Mathen and Chakraborty procedure where it may be necessary to resort to the evaluation of a polynomial in p, the proportion of infected individuals. Moreover, the calculation of the mean and variance of the Mathen and Chakraborty statistic with unequal household size is rather tedious.

Normal approximations may be used although a Poisson approximation is preferable when the proportion of infected individuals is small.

Unless a complete health survey is taken, the data will at best consist of details of households with at least one infected individual. We will assume that all infected households are included (complete reporting) and that multiply-infected households are included only once in the

data.  Households with no cases are not included.  To obtain
the mean and expected values of the resulting statistic,
substitute (n-s) for n, $(m_i-1)$ for $m_i$, and (r-s) for r in
(1) and (2).  Also, the following expressions are needed:

$$\sum_i (m_i-1)^2 = S_2 - 2n + s,$$

$$\sum_i (m_i-1)^3 = S_3 - 3S_2 + 3n - s .$$

The asymptotic distribution of $N_r$ is normal when $n \to \infty$ such
that $r/n \to p$ where p is the proportion of the population
infected.  If the infection rate is low or $p << 1$, then the
assumption of a Poisson distribution for $N_r$ may be
reasonable.

Walter has compared the power of his pairs test with
the test of Mathen and Chakraborty.  He considers a
continuous infection model [e.g. Bailey (1957)] as the
alternative hypothesis (assuming cases to be equally
infective from onset of the disease to the end of the study
period) and demonstrates that in situations he examines, the
pairs test has greater power.  However, the difference in
power was small over the range of alternatives simulated.
The power advantage of the Walter test as compared to the
Mathen and Chakraborty test is greater when there are more
cases observed and for higher values of the contagious
effect.

D.  Extensions of Mathen and Chakraborty and of Walter
Two statistical tests have been discussed for detecting
household aggregation of disease.  Mathen and Chakraborty

(1950) proposed counting the number of households, Z, in which there were no cases of the disease, and Walter (1974) proposed counting the number of pairs, N, of cases within households. As originally formulated, both of these methods suffer from the limitation that it is assumed in the null case that all members of the population who are assumed to be susceptible are at equal risk to disease. Smith and Pike (1976) generalize these tests to the situation in which different population strata are at different risks to disease. The revised Walter's test is shown to be a special case of a test of Pike and Smith (1968).

As just noted, both the method by Walter and the method by Mathen and Chakraborty assume in the null case that all members of the study population who are assumed to be susceptible are at equal risk to the disease. This assumption may be reasonable, for example, when an infectious disease such as influenza enters a population in which there is no prior immunity. However, for many diseases the attack rate is very much influenced by factors such as a person's age, sex, area of residence or immune status. In such situations, the assumption of equal susceptibility of all members of the population may produce a significant statistical test of aggregation of cases within households due solely to variation in the susceptibility status of members of different households.

Suppose we have a population of size t which is divided into h households, in which n cases of the disease under

study are observed (let m = t-n). The members of the
population are divided into k strata and it is assumed under
the null hypothesis that within each stratum all members are
at equal risk of the disease. Define $n_{ij}$ = number of cases
of disease and $m_{ij}$ = number of unaffected persons in the $j^{th}$
stratum of the $i^{th}$ household where $i = 1,...,h$ and
$j = 1,...,k$. Now

$$\Sigma n_{ij} = n_j, \quad n_{ij} + m_{ij} = t_{ij},$$
$$\Sigma m_{ij} = m_j, \quad n_j + m_j = t_j .$$

In the null case we assume that the $n_j$ cases in the $j^{th}$
stratum are drawn at random from the $t_j$ persons in that
stratum $(j=1,...,k)$. Under this randomization, the
probability that the $i^{th}$ household contains x cases is given
by

$$P_i(x) = \Sigma \prod_{j=1}^{k} \binom{n_j}{x_j} \binom{m_j}{t_{ij} - x_j} \Big/ \binom{t_j}{t_{ij}}$$

where $\Sigma$ extends over all $x_j$ such that $\sum_{j=1}^{k} x_j = x$ and

$x_j \le t_{ij}$, $x_j \le n_j$, $t_{ij} \le t_j$, and $x_j \ge 0$. The expected
number of households containing x cases is

$$E(x) = \sum_{i=1}^{h} P_i(x) .$$

We may thus compare the observed distribution of cases among
households with that expected. However, the usual chi-
square goodness of fit test may be a poor test of this
hypothesis as it is based on multiple degrees of freedom.
We therefore consider single degree of freedom tests based

upon the number of households with no cases and the number of pairs of cases within households.

## 1. *The number of households with no cases, Z*

The null distribution of Z is obtained by randomly allocating for each j the title of case to $n_j$ of the $t_j$ persons in the $j^{th}$ stratum. An approximate significance level may be obtained by simulation or by approximating the distribution of Z through its low order moments.

$$E(Z) = \sum_j \{\prod_j \prod_s [(m_j - s + 1) / (t_j - s + 1)]\},$$

$$E(Z_i Z_p) = \prod_j \prod_{s=1}^{t_{ij}+t_{pj}} [(m_j - s + 1) / (t_j - s + 1)],$$

$$V(Z) = E(Z) + 2 \sum_{i=1}^{h-1} \sum_{p=i+1}^{h} E(Z_i Z_p) - E(Z)^2.$$

Mathen and Chakraborty considered the case of households of equal size, say j, with all persons at equal risk of developing the disease. The above formulae reduce to their formulae in this case. For the case of unequal households sizes, these formulae simplify to formulae different from those given by Mathen and Chakraborty who considered a more restricted randomization scheme. They randomize such that $n_{(j)}$ cases observed among persons in households of size j were allocated only within households of size j.

## 2. *The number of pairs of cases within households, N*

The null distribution of N is obtained as for Z by randomly allocating for each j the title of case to $n_j$ of

the $t_j$ persons in the $j^{th}$ stratum.  An approximate
significance level may be obtained by simulation or by
approximating the distribution on N through its low order
moments.

Smith and Pike (1976) have generalized these tests for
familial aggregation even further.  Consider the problem of
comparing the amount of contact recorded among all possible
pairs of n patients with a disease with that recorded among
m controls, where the patients and controls are divided into
k strata, based upon factors which are likely to affect a
person's contact with others.  The $j^{th}$ stratum consists of
$n_j$ patients and $m_j$ controls.

Let $x_{bvjw}$ be the measure of contact from the $v^{th}$
patient in stratum b to the $w^{th}$ patient in stratum
j ($x_{bvbv} = 0$).  The sum of these measures over all possible
pairs of the n patients is

$$Z = \frac{1}{2} \sum_{b=1}^{k} \sum_{v=1}^{n_b} \sum_{j=1}^{k} \sum_{w=1}^{n_j} x_{bvjw}.$$

The null distribution of Z is derived by selecting as
patients n persons at random from the total of (m+n) cases
and controls, taking $n_j$ persons from the $(m_j+n_j)$ in the $j^{th}$
stratum (j=1,...,k).

Z can be related to N as follows:

(1) Equate the $n_j$'s for the two problems.

(2) Equate the above $m_j$ to the number of unaffected
persons in stratum j in the household aggregation problem.

(3) Let $x_{bvjw} = 1$ if the $v^{th}$ patient in stratum b and
the $w^{th}$ patient in stratum j come from the same household,

otherwise $x_{bvjw} = 0$. Under these definitions Z=N and the
randomization distributions are identical. Formulae for the
expectation and variance of Z and hence of N are given in
Pike and Smith (1968). Smith and Pike (1976) conjecture
that $[N - E(N)]^2/V(N)$ possibly with a continuity correction
may be taken as distributed approximately as chi-square on
one degree of freedom. It would seem reasonable to
speculate the Walter's power comparisons between his test
and the test of Mathen and Chakraborty (1950) would extend
directly to the more general case considered by Smith and
Pike (1976). On intuitive grounds, one would expect the
pairs test of have greater power as Mathen and Chakraborty's
(1950) test utilizes none of the information relating to the
distribution of cases among households with one or more
cases.

The methodology of Smith and Pike (1976) may have
useful application to situations other than the study of
familial aggregation such as occupational studies where the
work setting may be divided into somewhat separate work
areas.


E.  Applications

Bruckner *et al.* (1979) analyzed data collected over all the
endemic nephropathy cases during 1957-1976 in the endemic
village of Bistrita for time-space clustering and the
particular type of spatial clustering within household.
Knox's space-time interaction test and Walter's pair

statistic test were utilized. From r observed cases of the disease, the $r(r-1)/2$ pairs of cases were examined. For each distinct pair of cases i and j, two measures were obtained: $X_{ij}$ related to time separation between the two cases and $Y_{ij}$ related to spatial separation. The statistical test of the space-time interaction was based on

$$Z = \sum_i \sum_j X_{ij} Y_{ij} .$$

The value of Z in excess from its null expectation $E(Z)$ divided by standard deviation was used as evidence of space-time interaction. The graph intersection approach of Barton and David and Mantel's permutational variance procedure were used as computational methods to estimate these parameters.

Walter's pair statistic $N_r$ (see Section XI C) takes into account a particular type of spatial clustering within households. The null hypothesis assumes the lack of a person-to-person infection, or that all members of the population are equally prone to be infected. Evidence of the infectious nature of a disease can be provided by a significant deviation of $N_r$ from its null expectation. Recall that $N_r$ is defined as

$$N_r = \sum_{i=1}^{s} \frac{r_i(r_i-1)}{2}$$

where s is the number of households and $r_i$ the number of cases in the $i^{th}$ household. Evaluation of the first two moments of $N_r$ were made using the hypergeometric distribution of $r_i$.

Bruckner *et al.* (1979) found no evidence of time-space clustering of disease by Knox's procedure and no evidence of familial aggregation by Walter's test.


XII.  SUMMARY

Space-time clustering of disease is an interaction between the places of onset and the times of onset of the disease. Various methods for detecting clustering in both space and time have been developed.  Knox (1964 a) proposed a criterion based upon all possible pairs of cases.  The test statistic is the number of pairs close in time and close in space.  Knox (1964 b) conjectured that the distribution of this statistic was Poisson.  David and Barton (1966) gave the exact mean and variance of the statistic.  The main difficulty with Knox's method is the specification of critical times and distances for an unidentified disease process.  Knox's method was extended by Pike and Smith (1968) to take into account assumed periods of infectivity and susceptibility of cases.  Their statistic can be evaluated against its randomization distribution.  However, it is often necessary to adopt a Monte Carlo approach or to approximate its distribution.  Mantel (1967) generalized Knox's procedure for detecting time-space clustering of disease by defining the following statistic:

$$Z = \sum_{i<j} \sum X_{ij} Y_{ij}$$

where $X_{ij}$ is a spatial measure between points i and j and

$Y_{ij}$ is a temporal measure. Mantel (1967) presents formulae
for the expectation of Z and its permutational variance.
Pike and Smith (1974) suggest an approach for investigating
time-space clustering of disease using information derived
from a set of matched control persons as well as the cases
themselves. This case-control approach attempts to
ascertain whether patients with the disease have had more
relevant contact with each other than have a suitably chosen
control group. Mantel's generalized regression approach has
been extended to the two sample case by Klauber (1971). In
the two-sample situation one is confronted with two sets A
and B of cases having space coordinates and time
coordinates. One may wish to detect whether or not there is
a tendency for pairs of points, one from A and one from B,
which are spatially close to be also temporally close and
conversely. The two-sample randomization tests for space-
time clustering of Klauber (1971) have been extended to
cases with more than two samples by Klauber (1975).

Somewhat different approaches to the problem of testing
for time-space clustering of disease have been presented by
Pinkel *et al*. (1963), Ederer *et al*. (1964), and David and
Barton (1966). Of these methods, the one proposed by Ederer
*et al*. (1964) based upon occupancy numbers has been the most
popular. With any of these methods, whenever there is
significant evidence of clustering, one must consider the
possibility of artifactual causes. Many of these methods
have been used in the cancer literature especially for such

diseases as Burkitt's lymphoma, Hodgkin's disease, and leukemia. The strongest evidence for clustering is for Burkitt's lymphoma in Africa. Much of this literature has been recently reviewed by Smith (1978).

Various methods for detecting time-space clustering of disease have been applied to the problem of testing for familial aggregation. Walter's test (1974) is a special formulation of the test proposed by Knox (1964). Smith and Pike (1976) have generalized Walter's test to the situation in which different population strata are at different risks to the disease. Bruckner (1979) has recently applied these techniques to the study of familial aggregation in endemic nephropathy.

Continuing methodological research in this area is represented by the recent work of Mielke (1978), Siemiatycki (1978), and Wallenstein (1980).

## REFERENCES

Abe, O. (1973). A note on the methodology of Knox's tests of "time and space interaction". *Biometrics*, 29: 68-77.

Alderson, M.R. and Nayak, R. (1971). A study of space-time clustering in Hodgkin's disease in the Manchester region. *British Journal of Preventive and Social Medicine*, 25: 168-173.

Bailey, N.T.J. (1957). *The Mathematical Theory of Epidemics*. Griffin, London.

Barton, D.E., David, F.N., and Merrington, M. (1965). A criterion for testing contagion in time and space. *Annals of Human Genetics*, 29: 97-102.

Barton, D.E. and David, F.N. (1966). The random
    intersection of two graphs. In: *Research Papers in
    Statistics*. F.N. David (ed.). Wiley, New York,
    pp. 455-459.

Bross, I.D.J. (1958). How to use ridit analysis.
    *Biometrics*, 14: 18-38.

Brubaker, G., Geser, A., and Pike, M.C. (1973). Burkitt's
    lymphoma in the North Main District of Tanzania
    1964-1970: Failure to find evidence of time-space
    clustering in a high risk isolated rural area.
    *British Journal of Cancer*, 28: 469-472.

Bruckner, I., Rusu, G., Nichifor, E., and Giurcaneanu,
    C. (1979). Application of mathematical methods to
    the study of familial aggregation in endemic
    nephropathy I. The hypothesis of infections
    etiology. *Medecine Interne*, 17: 125-129.

David, F.N. and Barton, D.E. (1966). Two space-time
    interaction tests for epidemicity. *British Journal
    of Preventive and Social Medicine*, 20: 44-48.

Doll, R. (1978). An epidemiological perspective of the
    biology of cancer. *Cancer Research*, 38: 3573-3583.

Ederer, F., Meyers, M.H., and Mantel, N. (1964). A
    statistical problem in space and time: Do leukemia
    cases come in clusters? *Biometrics*, 20: 626-638.

Ederer, F., Myers, M.H., Eisenberg, H., and Campbell,
    P.C. (1965). Temporal-spatial distribution of
    leukemia and lymphoma in Connecticut. *Journal of
    the National Cancer Institute*, 35: 625-629.

Feller, W. (1957). *An Introduction to Probability Theory
    and its Application*. Volume 1, 2nd ed. Wiley, New
    York.

Glass, A.G. and Mantel, N. (1969). Lack of space-time
    clustering of childhood leukemia, Los Angeles
    County, 1960-64. *Cancer Research*, 29: 1995.

Hamadeh, R.R., Armenian, H.K., and Zurayk, H.C. (1980). A
    study of clustering of cases of leukemia, Hodgkin's
    disease and other lymphoma's in Bahrain. *Tropical
    and Geographic Medicine*, 33: 42-48.

Heath,C.W., Jr., and Hasterlik, R.J. (1963). Leukemia among
    children in a suburban community. *American Journal
    of Medicine*, 34: 796-812.

Heath, A.B. (1977).    Cleft lip and palate in the Oxford
    area:  An examination of the evidence for clustering
    in space and time.  *British Journal of Preventive
    and Social Medicine* 31: 269-271.

Hoeffding, W. (1948).    A class of statistics with
    asymptotically normal distribution.  *Annals of
    Mathematical Statistics*, 19: 293-325.

Klauber, M.R. (1971).    Two-sample randomization tests for
    space-time clustering.  *Biometrics*, 27: 129-142.

Klauber, M.R. (1975).    Space-time clustering tests for more
    than two samples.  *Biometrics*, 31: 719-726.

Klauber, R. and Mustacch, P. (1970).    Space-time clustering
    of childhood leukemia in San Francisco.  *Cancer
    Research*, 30: 1969-1973.

Knox, G. (1963).    Detection of low intensity epidemicity:
    application to cleft lip and palate.  *British
    Journal of Preventive and Social Medicine*, 17:
    121-127.

Knox, G. (1964 a).    Epidemiology of childhood leukemia in
    Northumberland and Durham.  *British Journal of
    Preventive and Social Medicine*, 18: 17-24.

Knox, G. (1964 b).    The detection of space-time
    interactions.  *Applied Statistics*, 13: 25-29.

Kryscio, R.J., Myers, M.H., Presiner, S.T., et al. (1973).
    The space-time distribution of Hodgkin's disease in
    Connecticut, 1940-1969.  *Journal of the National
    Cancer Institute*, 50: 1107-1110.

Lehmann, E.L. (1951).    Consistency and unbiasedness of
    certain nonparametric tests.  *Annals of Mathematical
    Statistics*, 22: 165-179.

Lloyd, S. and Roberts, C.H. (1977).    A test for space
    clustering and its application to congenital limb
    defects in Cardiff.  *British Journal of Preventive
    and Social Medicine*, 27: 188-191.

Mantel, N. (1967).    The detection of disease clustering and
    a generalized regression approach.  *Cancer Research*,
    27: 209-220.

Mantel, N. and Bailar, J.C. (1970).    A class of
    permutational and multinomial test arising in
    epidemiological research.  *Biometrics*, 26: 687-700.

Mantel, N. and Valand, R.S. (1970). A technique of non-parametric multivariate analysis. *Biometrics*, 26: 547-548.

Mantel, N., Kryscio, R.J., and Myers, M.H. (1976). Tables and formulas for extended use of the Ederer-Myers-Mantel disease-clustering procedure. *American Journal of Epidemiology*, 104: 576-584.

Mathen, K.K. and Chakraborty, P.N. (1950). A statistical study on multiple cases of disease in households. *Sankhya*, 10: 387-392.

Mielke, P.W., Berry, K.J., and Johnson, E.S. (1976). Multi-response permutation procedures for a priori classifications. *Communications in Statistics, Theory and Methods*, 5A: 1409-1424.

Mielke, P.W. (1978). Clarification and appropriate inferences for Mantel and Valand's nonparametric multivariate analysis technique. *Biometrics*, 34: 277-282.

Morrow, R.H., Pike, M.C., Smith, P.G., Ziegler, J.L., and Kissuule, A. (1970). Burkitt's lymphoma: A time-space cluster of cases in Bwamba County of Uganda. *British Medical Journal*, 2: 491-492.

Morrow, R.H., Pike, M.C., and Smith, P.G. (1977). Further studies of space-time clustering of Burkitt's lymphoma in Uganda. *British Journal of Cancer*, 35: 668-673.

Naus, J. (1966). Some probabilities, expectations, and variances for the size of smallest intervals and largest clusters. *Journal of American Statistical Association*, 61: 1191-1199.

Pike, M.C. and Smith, P.G. (1968). Disease clustering: A generalization of Knox's approach to the detection of space-time interactions. *Biometrics*, 24: 541-556.

Pike, M.C. and Smith, P.G. (1974). A case-control approach to examine diseases for evidence of contagion, including diseases with long latent periods. *Biometrics*, 30: 263-279.

Pinkel, K., Dawd, J.E., and Bross, I.D.J. (1963). Some epidemiological features of malignant solid tumors of children in the Buffalo, N.Y. area. *Cancer*, 16: 28-33.

Pinkel, D. and Nefzger, D. (1959). Some epidemiological features of childhood leukemia in the Buffalo, N.Y. area. *Cancer, 12*: 351-358.

Pleydell, M.J. (1957). Mongolism and other congenital abnormalities: Epidemiological study in Northamptionshire. *Lancet, 272*: 1314-1319.

Poskanzer, D.C., Walker, A.M., Prenney, L.B., and Sheridan, J.L. (1981). The etiology of multiple sclerosis: Temporal-spatial clustering indicating two environmental exposures before onset. *Neurology, 31*: 708-713.

Roberts, C.H., Laurence, K.M., and Lloyd, S. (1975). An investigation of space and space-time clustering in a large sample of infants with neural tube defects born in Cardiff. *British Journal of Preventive and Social Medicine, 29*: 202-204.

Siemiatycki, J. (1971). Space-time Clustering: Finding the Distribution of a Correlation-Type Statistic. M.Sc. Thesis, McGill University.

Siemiatycki, J. (1978). Mantel's space-time clustering statistic: computing higher moments and a comparison of various data transforms. *Journal of Statistical Computing and Simulation, 7*: 13-31.

Siemiatycki, J., Brubaker, G., and Geser, A. (1980). Space-time clustering of Burkitt's lymphoma in East Africa: Analysis of recent data and a new look at old data. *British Journal of Cancer, 25*: 197-203.

Smith, P.G., Pike, M.C., Till, M.M., and Hardesty, R.M. (1976). Epidemiology of childhood leukemia in Greater London: a search for evidence of transmissions assuming a possible long latent period. *British Journal of Cancer, 33*: 1-8.

Smith, P.G. and Pike, M.C. (1976). Generalizations of two tests for the detection of household aggregation of disease. *Biometrics, 32*: 817-828.

Smith, P.G., Pike, M.C., Inlen, L.J., Jones, A., and Harris R. (1977). Contacts between young patients with Hodgkin's disease. *Lancet, 2*: 59-62.

Smith, P.G. (1978). Current assessment of 'case clustering' of lymphomas and leukemias. *Cancer, 42*: 1026-1034.

Snow, J. (1855). *On the Mode of Communication of Cholera.* Churchill, London.

Stark, C.R. and Mantel, N. (1967). Lack of seasonal or
    temporal-spatial clustering of Down's syndrome
    births in Michigan. *American Journal of
    Epidemiology*, *86*: 199-213.

Stark, C.R. and Mantel, N. (1967). Temporal-spatial
    distribution of birth dates for Michigan children
    with leukemia. *Cancer Research*, *27*: 1749-1775.

Vianna, N.J., Greenwald, P., and Davis, J.N.P. (1971).
    Extended epidemic of Hodgkin's disease in high-
    school students. *Lancet*, *i*, 1209-1211.

Vianna, N.J., Greenwald, P., Brady, J., *et al.* (1972).
    Hodgkin's disease: Cases with features of a
    community outbreak. *Annals of Internal Medicine*,
    *77*: 169-180.

Wallenstein, S. (1980). A test for detection of clustering
    over time. *American Journal of Epidemiology*, *111*:
    367-372.

Walter, S.D. (1974). On the detection of household
    aggregation of disease. *Biometrics*, *30*: 525-538.

Williams, E.H., Smith, P.G., Day, N.E., Geser, A., Ellice,
    J., and Tukei, P. (1978). Space-time clustering of
    Burkitt's lymphoma in the West Nile District of
    Uganda: 1961-1975. *British Journal of Cancer*, *37*:
    109-122.

# 6
# STATISTICAL METHODS FOR GENETIC STUDIES
# OF HLA AND CANCER

John J. Gart
National Cancer Institute
Chevy Chase, Maryland

Jun-mo Nam
National Cancer Institute
Bethesda, Maryland

## I. INTRODUCTION

The genesis of the human leukocyte antigen (HLA) system is
the need for a scientific basis for matching donors and
recipients in transplantation. It is the human analogue of
the mouse histocompatibility system, H-2. As this genetic
system has a major effect on the survival of a
transplantable tumor and the susceptibility to viral
leukemogenesis in mice, it is conjectured that the HLA may
provide clues to cancer in humans.

Several case-control studies have investigated the
possible association of various HLA antigens and cancer.
Various authors (see Ryder and Svejgaard (1976)) have

implicated retinoblastoma, acute lymphocytic leukemia (ALL),
and nasopharyngeal carcinoma (NPC).  Non-malignant diseases,
for instance ankylosing spondylitis, have also been shown to
be associated with the HLA system.

This paper considers the statistical methodology
appropriate for analyzing such studies.  Estimation and
testing of antigen and gene frequencies are reviewed or
developed.  In particular, omnibus tests which compare all
the antigen or gene frequencies at a given locus are
developed.  The methods are illustrated using case-control
studies on ALL and NPC.

A.  Description of HLA System and Possible Structures of the
    Data

The HLA system is a generalized ABO-like system in which
there are more than two alleles.  It consists of several
loci.  By 1980, the loci A (or LA series), B (or series 4),
C, D, and DR are recognized.  (See Terasaki (1980)).
Associated with the first locus there are several codominant
alleles (or antigens), say $A_1, A_2, \ldots, A_{m-1}$, and a recessive
allele 0, which maybe considered as a pool of antigens not
yet identified.  Similar sets of several alleles are
associated with the B, C, D, and DR loci.  An individual may
possess 2, 1, or none of the alleles at a given locus, for
instance, at the A locus, the possible types are:  $A_i A_i$,
$i = 1, 2, \ldots, m-1$, $A_i A_j$, $i \neq j = 1, 2, \ldots, m-1$, $A_i 0$, $i = 1, 2, \ldots, m-1$,

and 00. This totals $m(m+1)/2$ distinct genotypes. Unless typing of other family members is done, it is not possible to distinguish individuals of genotype, $A_iA_i$, from individuals of genotype, $A_i0$. These individuals are denoted $A_i$, $i=1,2,\ldots,m-1$. Such data, called phenotypic data, have only $(m^2-m+2)/2$ distinct phenotypes, and in the terminology of Cotterman (1953), this is called the phenogram, $m-(m^2-m+2)/2-1$; it has m total alleles, $(m^2-m+2)/2$ phenotypes, and one recessive allele. In a case-control study, one of these two types of data is found for a set of $N_1$ cases and $N_2$ controls. It is convenient to array such data in the form of matrices. This is illustrated in Table 1. Note that $n_{ij}=n_{ji}$ for all i, j. Many authors only report the antigen frequencies: $G_i=n_i+\Sigma_j n_{ij}$, for $j\neq i=1,2,\ldots,m-1$ and $n_{oo}$, for phenotypic data, or

$$G_i = \sum_{j=1}^{m-1} n_{ij} + n_{io}$$

and $n_{oo}$ for genotypic data. Unfortunately such presentation may result in some needless loss of efficiency in estimation and loss of power in statistical testing. In Section II we present methods appropriate for analyses of such data in the literature. It should be pointed out however that the $n_{ij}$'s, $n_{io}$'s, $n_{ii}$'s, or $n_i$'s, and $n_{oo}$ all will be available if the individual data were properly recorded. That is, no additional biological testing is needed, simply more careful record keeping.

Table 1    The Structure and Notation for the Various
           Types of Data

| Antigens | Phenotypic data | | | | $\Sigma$ |
| --- | --- | --- | --- | --- | --- |
| | $A_1$ | $A_2$ | $\cdots$ | $A_{m-1}$ | $0$ | |
| $A_1$ | $n_1$ | $n_{12}$ | $\cdots$ | $n_{1,m-1}$ | $-$ | $G_1$ |
| $A_2$ | $n_{21}$ | $n_2$ | $\cdots$ | $n_{2,m-1}$ | $-$ | $G_2$ |
| . | . | . | | . | . | . |
| . | . | . | | . | . | . |
| . | . | . | | . | . | . |
| $A_{m-1}$ | $n_{m-1,1}$ | $n_{m-1,2}$ | $\cdots$ | $n_{m-1}$ | $-$ | $G_{m-1}$ |
| $0$ | $-$ | $-$ | $\cdots$ | $-$ | $n_{oo}$ | $-$ |

| Antigens | Genotypic data | | | | | $\Sigma$ |
| --- | --- | --- | --- | --- | --- | --- |
| | $A_1$ | $A_2$ | $\cdots$ | $A_{m-1}$ | $0$ | |
| $A_1$ | $n_{11}$ | $n_{12}$ | $\cdots$ | $n_{1,m-1}$ | $n_{1o}$ | $G_1$ |
| $A_2$ | $n_{21}$ | $n_{22}$ | $\cdots$ | $n_{2,m-1}$ | $n_{2o}$ | $G_2$ |
| . | . | . | | . | . | . |
| . | . | . | | . | . | . |
| . | . | . | | . | . | . |
| $A_{m-1}$ | $n_{m-1,1}$ | $n_{m-1,2}$ | $\cdots$ | $n_{m-1,m-1}$ | $n_{m-1,o}$ | $G_{m-1}$ |
| $0$ | $n_{o1}$ | $n_{o2}$ | $\cdots$ | $n_{o,m-1}$ | $n_{oo}$ | $n_{o.}$ |

## B. The Hardy-Weinberg Model

The Hardy-Weinberg or random mating (H-W) model assumes that the groups of size N follow a multinomial distribution with genotypic probabilities (Yasuda and Kimura (1968)):

$$P(A_i A_j) = 2p_i p_j, \qquad i \neq j = 1, 2, \ldots, m-1,$$
$$P(00) = r^2,$$

and
$$P(A_i 0) = 2p_i r,$$
$$P(A_i A_i) = p_i^2, \qquad i = 1, 2, \ldots, m-1.$$

For phenotypic frequencies, the $A_i A_i$ and $A_i 0$ are merged with the probabilities:

$$P(A_i) = p_i^2 + 2p_i r, \qquad i = 1, 2, \ldots, m-1,$$

but otherwise it is the same as the genotypic frequencies for $P(A_i A_j)$ and $P(00)$. The $p_i$'s and r are the gene or allele frequencies, where $r > 0$, $p_i > 0$, $i = 1, 2, \ldots, m-1$, and $\Sigma_i p_i + r = 1$. If the Hardy-Weinberg law holds, the most efficient comparison of case with control group is done by testing differences between the respective vectors of gene frequencies $\underline{p}' = (p_1, p_2, \ldots, p_{m-1})$.

Note that when m=3, all of the above model and notation reduces to that for the ABO system.

## II. STATISTICAL ANALYSES OF ANTIGEN FREQUENCIES

In this section we consider the statistical analyses of studies in which only the antigen frequencies, $G_i$, $i = 1, 2, \ldots, m-1$, and the double-blank frequency, $n_{oo}$, are available from a total of N individuals (see, e.g., data of

Simons, Wee, Goh, Chan, Shanmugaratnam, Day and de-Thé
(1976) given later in Table 2). The gene frequency
estimators, a test of fit of the H-W model, and tests of
differences between the case and control groups are
discussed.

A.  Gene Frequency Estimation and Fitting of the H-W Model

Under the assumption of the H-W model it is easily shown
that the "expected" frequencies of the data given in this
form are:

$$P_i = E(G_i/N) = 1-(1 - p_i)^2, \quad i=1,2,\ldots,m-1, \qquad (1)$$

and

$$P_{oo} = E(n_{oo}/N) = r^2.$$

Equating observed and expected frequencies, we easily find
the simple Bernstein's estimators of the gene frequencies
(Yasuda and Kimura (1968)):

$$\hat{p}_i = 1-(1 - G_i/N)^{1/2}, \qquad i=1,2,\ldots,m-1, \qquad (2)$$

and

$$\hat{r} = [(n_{oo}/N)]^{1/2}.$$

Note that the estimated gene frequencies do not necessarily
sum to one, and denote this discrepancy by

$$D = 1-(\Sigma\hat{p}_i + \hat{r}),$$

which may be positive or negative. The simple Bernstein's estimators are generally inefficient, unless m=2, in which case they are the maximum likelihood estimators (MLE's) and are thus fully efficient. Stevens (1950) suggested that the discrepancy noted above may be reduced by using the so-called "adjusted" Bernstein estimators. These are fully efficient for m=3 (i.e., the ABO system). Nam and Gart (1976) suggested alternative "modified" Bernstein's estimators:

$$\hat{p}_i' = \hat{p}_i / (1-D/2),$$

and                                                                    (3)

$$\hat{r}' = (\hat{r} + D/2)/(1 - D/2).$$

An obvious amicable property of these estimators is that $\Sigma_i \hat{p}_i' + \hat{r}' = 1$ all the time, and these estimators have the same asymptotic properties as the "adjusted" method. Unfortunately it is still possible, as in the adjusted method, for $\hat{r}'$ to be negative.

Nam and Gart showed that for m≥4, the adjusted estimators are inefficient. In some instances, for $m \geq 4$ and r large, the adjusted (or the equivalent modified) estimators are not only inefficient, but may have asymptotic variances larger than the simple estimators. In the typical case when r is small the modified estimators may represent an improvement over the simple estimators. However, their standard errors are extremely messy to compute (see equations (6.3) and (6.4) of Nam and Gart (1976)).

The asymptotic variance-covariance matrix of the simple Bernstein's estimators are easily derived (see e.g. Nam and Gart (1976)):

$$V(\hat{p}_i) = p_i(2-p_i)/(4N), \qquad\qquad i=1,2,\ldots,m-1,$$

$$C(\hat{p}_i,\hat{p}_j) = -\frac{p_i p_j}{2N}\left\{1 - \frac{p_i p_j}{2(1-p_i)(1-p_j)}\right\}, \quad i\neq j=1,2,\ldots,m-1,$$

$$V(\hat{r}) = (1-r^2)/(4N), \qquad \text{for } r > 0,$$

and                                                                                  (4)

$$C(\hat{p}_i,\hat{r}) = -\frac{rp_i}{4N}\left\{1 + \frac{1}{(1-p_i)}\right\}, \quad i=1,2,\ldots,m-1.$$

As well as providing estimated standard errors for the estimators, these results can be used to provide a simple test of the goodness of fit of the H-W model. Note that under the H-W model, $E(D)=0$ asymptotically. Using the results above, Nam and Gart (1976) have shown asymptotically that

$$V(D) = \frac{1}{4N}\left\{\left(\sum\frac{p_i}{1-p_i}\right)^2 - \sum\left(\frac{p_i}{1-p_i}\right)^2\right\}.$$     (5)

Thus an approximately chi-square test with a single degree of freedom is

$$\chi_D^2 = D^2/\hat{V}(D),$$                                                        (6)

where $\hat{V}(D)$ is (5) with $p_i=\hat{p}_i$. When m=3, (6) reduces to the test of fit given by Stevens (1950, equation 2.24) for the ABO system.

B.  Testing for Differences in Individual Antigen or Gene
Frequencies

Most authors test for association of antigen, $A_i$, with
disease by comparing the case group antigen frequency,
$\hat{P}_{i1} = G_{i1}/N_1$, with control group antigen frequency,
$\hat{P}_{i2} = G_{i2}/N_2$, by the usual one degree of freedom chi square
test of a two by two table:

|         | $A_i$     | Non $A_i$       | Sums  |
|---------|-----------|-----------------|-------|
| Case    | $G_{i1}$  | $N_1-G_{i1}$    | $N_1$ |
| Control | $G_{i2}$  | $N_2-G_{i2}$    | $N_2$ |
| Sums    | $G_i\cdot$| $N\cdot-G_i\cdot$ | $N\cdot$ |

$$\chi^2_{Ai} = \frac{(\hat{P}_{i1}-\hat{P}_{i2})^2}{\hat{P}_{i\cdot}(1-\hat{P}_{i\cdot})(1/N_1 + 1/N_2)}, \tag{7}$$

where $\hat{P}_{i\cdot} = G_{i\cdot}/N\cdot$.  If a one-tailed test is required, the
normal deviate is $Z_{Ai} = (\chi^2_{Ai})^{1/2}$ with the appropriate
algebraic sign.

   It is of interest to investigate whether comparing the
gene frequency estimators will lead to a more efficient
test.  The appropriate normal deviate for the simple
Bernstein's estimator is:
$$Z_{Gi} = (\hat{p}_{i1} - \hat{p}_{2i})/[\bar{p}_i(2-\bar{p}_i)\{1/(4N_1) + 1/(4N_2)\}]^{1/2}$$
where $\bar{p}_i$ is estimated from the pooled sample,
$$\bar{p}_i = 1 - \{1 - (G_{i\cdot}/N\cdot)\}^{1/2}. \tag{8}$$
Using the concept of Pitman's relative efficiency

(e.g. Noether (1955)), one can easily show that $Z_{Ai}$ and $Z_{Gi}$ have a relative efficiency of one. This is intuitively clear from the fact that each statistic is a function of the $G_{i1}$ and $G_{i2}$ alone. Thus nothing is gained from invoking the H-W model and using Bernstein's simple estimators.

A similar normal deviate test may be constructed using the modified Bernstein's method with their appropriate asymptotic variance formulas. The relative efficiency of such a test is the same as the relative efficiency of the simple Bernstein's estimator to the modified Bernstein's estimator. As noted above either method may be better depending on the values of the parameters. Nam and Gart (1976, p. 365 and Table 2) show specific examples in which one method or the other may be more efficient. The typical HLA frequencies are such that the considerable additional computing in estimating the variance of the modified estimators does not appear worth the effort. Nam and Gart (1976, p. 369) show that the simple Bernstein's estimator of the HL-A2 gene frequency has a relative efficiency to the modified estimator of 99.85%. The relative efficiencies for the simple Bernstein's estimators of the other gene frequencies are even larger. Additionally it should be noted that the test based on $Z_{Ai}$ is not dependent on the assumption of the H-W model.

A more important statistical question is that of making individual significance tests for each of the several gene or antigen frequencies. The question arises: If one tests

the differences of ten or fifteen gene frequencies between
the case and control groups, is it appropriate to report the
extreme P-value as if it were the only statistical test
performed?  Two situations can be delineated:

(1)  The study in question is being done to test a
hypothesis of a particular antigen having an increased
frequency in the case group over the control group.  For
instance, a second study may be done to confirm a previous
report in the literature about the association of a
particular antigen with a particular disease.  In this case
it is appropriate to use the one-tailed P-value from the
individual test of this antigen.

(2)  There is no prior hypothesis about the particular
antigen and there are m comparisons of antigen frequencies.
In this case the Bonferroni correction (Wilks (1962,
p. 290)) maybe used; that is, the m x P-value rather than
the P-value is compared to the usual significance level
(0.05 or 0.01).  Two-tailed, rather than one-tailed, tests
are usually also appropriate in this situation.  An
additional useful method in the latter situation to compare
all the gene frequencies at a given locus in the two groups
simultaneously in a single, omnibus test.  This is
considered in the next section.

C.  Omnibus Test of Antigen Frequency Differences

Since the antigen frequencies are sums of independent
variables it is clear that they have an asymptotic

Table 2  Analysis of Case-Control Study of NPC and HLA (Simons *et al.* (1976), First or LA Locus)

| Antigen | Cases ($N_1$=110) | | Controls ($N_2$=91) | | Pooled ($N_.$=201) | | |
|---|---|---|---|---|---|---|---|
| | $G_{i1}$ | $\hat{P}_{i1}$ | $G_{i2}$ | $\hat{P}_{i2}$ | $\hat{P}_{i.}$ | $\hat{P}_{i1}-\hat{P}_{i2}$ | $X^2_{Ai}$ |
| HLA – A1 | 1 | 0.0091 | 1 | 0.0110 | 0.0100 | -0.0019 | 0.018 |
| A2 | 66 | 0.6000 | 47 | 0.5165 | 0.5622 | 0.0835 | 1.411 |
| A3 | 2 | 0.0182 | 2 | 0.0220 | 0.0199 | -0.0038 | 0.037 |
| A9 | 41 | 0.3727 | 32 | 0.3516 | 0.3632 | 0.0211 | 0.096 |
| A10 | 7 | 0.0636 | 10 | 0.1099 | 0.0846 | -0.0463 | 1.376 |
| A11 | 50 | 0.4545 | 54 | 0.5934 | 0.5174 | -0.1389 | 3.846[1] |
| "AW19" | 21 | 0.1909 | 12 | 0.1319 | 0.1642 | 0.0590 | 1.265 |
| Double Blank ($n_{oo}$) | 0 | -- | 0 | -- | -- | -- | -- |

Omnibus test cases *vs.* control: $X^2_{A.}$ = 6.7103 (7 d.f.), P = 0.46

Table 2 (Continued)

| Bernstein estimates | Gene frequencies | | |
|---|---|---|---|
| | Cases ($N_1 = 110$) | Controls ($N_2 = 91$) | Pooled ($N. = 201$) |
| $\hat{p}_1$ | $0.0046 \pm 0.0046$ | $0.0055 \pm 0.0055$ | $0.0050 \pm 0.0035$ |
| $\hat{p}_2$ | $0.3675 \pm 0.0369$ | $0.3046 \pm 0.0377$ | $0.3383 \pm 0.0264$ |
| $\hat{p}_3$ | $0.0091 \pm 0.0064$ | $0.0111 \pm 0.0078$ | $0.0100 \pm 0.0050$ |
| $\hat{p}_4$ | $0.2080 \pm 0.0291$ | $0.1948 \pm 0.0311$ | $0.2020 \pm 0.0212$ |
| $\hat{p}_5$ | $0.0323 \pm 0.0120$ | $0.0565 \pm 0.0174$ | $0.0432 \pm 0.0102$ |
| $\hat{p}_6$ | $0.2615 \pm 0.0321$ | $0.3624 \pm 0.0404$ | $0.3053 \pm 0.0254$ |
| $\hat{p}_7$ | $0.1005 \pm 0.0208$ | $0.0683 \pm 0.0190$ | $0.0858 \pm 0.0143$ |
| $\hat{r}$ | $0.0000 \pm 0.0000$ | $0.0000 \pm 0.0000$ | $0.0000 \pm 0.0000$ |
| $\sum$ | $0.9835$ | $1.0032$ | $0.9896$ |
| Departure D from H-W Law | $0.0165$ | $-0.0032$ | $0.0104$ |
| SE(D) | $0.0542$ | $0.0614$ | $0.0404$ |
| Test of departure $\chi^2_D$ (1 d.f.) | $0.093$ | $0.003$ | $0.066$ |

[1] P-value $\simeq 0.05^-$ (two tailed)

multivariate normal distribution. Thus it is possible to
construct a test statistic which is a quadratic form of the
differences in the group's antigen frequencies and has an
asymptotic chi-square distribution. In formal terms we wish
to test the hypothesis $H_o$: $P_{i1} = P_{i2} = P_i$ (unknown) for
$i = 1,2,\ldots,m-1$ against $H_1 : P_{i1} \neq P_{i2}$ for one or more
$i = 1,2,\ldots,m-1$. To do this we consider the differences

$$d_i = G_{i1}/N_1 - G_{2i}/N_2, \qquad i=1,2,\ldots,m-1.$$

Define the vector, $\underline{d}' = [d_1,d_2,\ldots,d_{m-1}]$. Under the null
hypothesis, it is easy to show (from Nam and Gart (1976,
p. 367)) that

$$V(d_i) = p_i(2-p_i)(1-p_i)^2 \{1/N_1+1/N_2\}, \qquad i=1,2,\ldots,m-1,$$
and

$$C(d_i,d_j)=-p_i p_j\{2-2(p_i+p_j)+p_i p_j\}\{1/N_1+1/N_2\}, \quad i\neq j=1,2,\ldots,m-1$$

Recalling the fact that $E(G_{i.}/N.)$ can be written as
$p_i(2-p_i)$, we may estimate the variances and covariances by

$$v_{ii} = (G_{i.}/N.)(1-G_{i.}/N.)(1/N_1+1/N_2), \qquad i=1,2,\ldots,m-1,$$
and

$$v_{ij}= -(G_{i.}G_{j.}/N.^2-2\bar{p}_i\bar{p}_j)(1/N_1+1/N_2), \qquad i\neq j=1,2,\ldots,m-1,$$

where $\bar{p}_i$ is given by (8). Define $\underline{V} = [v_{ij}]$ to be the
corresponding symmetric $(m-1) \times (m-1)$ matrix. The
appropriate chi-square test statistic with m-1 degrees of
freedom is

$$\chi^2_{A.} = \underline{d}'\underline{V}^{-1}\underline{d}. \tag{9}$$

One could equally well form a test statistic from a
quadratic form of the vector of differences in simple
Bernstein estimators. Such a test will be asymptotically
equivalent to $\chi^2_{A.}$. A quadratic form of the differences in

modified Bernstein estimators could also be used as a test statistic. Since the total efficiencies of the simple and modified estimators are not all that different for HLA frequencies (Nam and Gart, 1976), it is not thought that the considerable additional computing involved is worth the effort.

It should be pointed out that it is not correct to sum the individual antigen frequency chi-squares (7) and treat the total as having a chi-square distribution with m-1 degrees of freedom (see, e.g., Svejgaard, Platz, Ryder, Nielsen, and Thomsen (1975)). This is, of course, due to the fact that these chi-squares are not independent variates, and thus their sum is not necessarily distributed as a chi-square variate.

D.  Antigen Frequency Analyses of NPC Data

Simons *et al.* (1976) did a case-control study of the HLA and nasopharyngeal carcinoma among Chinese in Singapore. They typed individuals on 7 antigens in the A locus (m=8) and 17 antigens in the B locus (m=18). They reported only antigen frequencies. There were 91 newly diagnosed cases and 110 controls. The data and results are given in Tables 2 and 3. Note that B14 and BW21 are omitted from these results because neither antigen is represented in the two groups. Thus m is reduced to 16 for the second locus. At the A locus there is only one difference, that of A11, which barely reaches significance at the 0.05 level by the chi-

Table 3   Analysis of Case Control Study of NPC and HLA (Simons *et al.* (1976), Second or Series 4 Locus)

| Antigen | Antigen frequencies | | | | | | |
|---|---|---|---|---|---|---|---|
| | Cases ($N_1 = 110$) | | Controls ($N_2 = 91$) | | $\hat{P}_{i.}$ | $\hat{P}_{i1} - \hat{P}_{i2}$ | $X^2_{Ai}$ |
| | $G_{i1}$ | $\hat{P}_{i1}$ | $G_{i2}$ | $\hat{P}_{i2}$ | | | |
| HLA-B5 | 15 | 0.1364 | 10 | 0.1099 | 0.1244 | 0.0265 | 0.320 |
| B7 | 2 | 0.0182 | 3 | 0.0330 | 0.0249 | -0.0148 | 0.449 |
| B8 | 2 | 0.0182 | 0 | 0.0000 | 0.0100 | 0.0182 | 1.671 |
| B12 | 3 | 0.0273 | 2 | 0.0220 | 0.0249 | 0.0053 | 0.058 |
| B13 | 19 | 0.1727 | 14 | 0.1538 | 0.1642 | 0.0189 | 0.129 |
| B18 | 2 | 0.0182 | 1 | 0.0110 | 0.0149 | 0.0072 | 0.175 |
| B27 | 1 | 0.0091 | 6 | 0.0659 | 0.0348 | -0.0568 | 4.787[1] |
| BW15 | 23 | 0.2091 | 22 | 0.2418 | 0.2239 | -0.0327 | 0.306 |
| BW16 | 13 | 0.1182 | 12 | 0.1319 | 0.1244 | -0.0137 | 0.086 |
| BW17 | 24 | 0.2182 | 24 | 0.2637 | 0.2388 | -0.0456 | 0.569 |
| BW22 | 5 | 0.0455 | 10 | 0.1099 | 0.0746 | -0.0644 | 2.994 |
| BW35 | 7 | 0.0636 | 7 | 0.0769 | 0.0697 | -0.0133 | 0.136 |
| BW37 | 1 | 0.0091 | 1 | 0.0110 | 0.0100 | -0.0019 | 0.018 |
| BW40 | 39 | 0.3545 | 35 | 0.3846 | 0.3682 | -0.0301 | 0.194[1] |
| SIN2 | 44 | 0.4000 | 23 | 0.2527 | 0.3333 | 0.1473 | 4.863[1] |
| Double blank | 0 | -- | 0 | -- | -- | -- | -- |

Omnibus test   cases *vs.* controls: $X^2_{A.} = 16.13$ (15 d.f.), P=0.37.

Table 3 (continued)

| Bernstein's estimates | Gene frequencies | | |
|---|---|---|---|
| | Cases ($N_1$ = 110) | Controls ($N_2$ = 91) | Pooled ($N.$ = 201) |
| $\hat{p}_1$ | 0.0707 ± 0.0176 | 0.0565 ± 0.0174 | 0.0643 ± 0.0124 |
| $\hat{p}_2$ | 0.0091 ± 0.0064 | 0.0166 ± 0.0095 | 0.0125 ± 0.0056 |
| $\hat{p}_3$ | 0.0091 ± 0.0064 | 0.0000 ± 0.0000 | 0.0050 ± 0.0035 |
| $\hat{p}_4$ | 0.0137 ± 0.0079 | 0.0111 ± 0.0078 | 0.0125 ± 0.0055 |
| $\hat{p}_5$ | 0.0905 ± 0.0198 | 0.0801 ± 0.0206 | 0.0858 ± 0.0143 |
| $\hat{p}_6$ | 0.0091 ± 0.0064 | 0.0055 ± 0.0055 | 0.0075 ± 0.0043 |
| $\hat{p}_7$ | 0.0046 ± 0.0046 | 0.0335 ± 0.0134 | 0.0176 ± 0.0066 |
| $\hat{p}_8$ | 0.1107 ± 0.0218 | 0.1292 ± 0.0258 | 0.1190 ± 0.0167 |
| $\hat{p}_9$ | 0.0609 ± 0.0164 | 0.0683 ± 0.0190 | 0.0643 ± 0.0124 |
| $\hat{p}_{10}$ | 0.1158 ± 0.0223 | 0.1419 ± 0.0269 | 0.1275 ± 0.0172 |
| $\hat{p}_{11}$ | 0.0230 ± 0.0102 | 0.0565 ± 0.0174 | 0.0380 ± 0.0096 |
| $\hat{p}_{12}$ | 0.0323 ± 0.0120 | 0.0392 ± 0.0145 | 0.0355 ± 0.0093 |
| $\hat{p}_{13}$ | 0.0046 ± 0.0046 | 0.0055 ± 0.0055 | 0.0050 ± 0.0035 |
| $\hat{p}_{14}$ | 0.1966 ± 0.0284 | 0.2155 ± 0.0325 | 0.2051 ± 0.0214 |
| $\hat{p}_{15}$ | 0.2254 ± 0.0302 | 0.1356 ± 0.0264 | 0.1835 ± 0.0204 |
| $\hat{r}$ | 0.0000 ± 0.0000 | 0.0000 ± 0.0000 | 0.0000 ± 0.0000 |
| $\Sigma$ | 0.9761 | 0.9950 | 0.9831 |
| Departure D from H-W Law | 0.0239 ± 0.0500 | 0.0050 ± 0.0556 | 0.0169 ± 0.0371 |
| Test of departure $X_D^2$ (1 d.f.) | 0.229 | 0.008 | 0.208 |

[1] P-value ≈ $0.03^-$ (two-tailed).

square test of individual antigen frequencies (7). This was
not a difference suspected *a priori* and a Bonferroni
correction of multiplication of its P-value by 8 renders it
insignificant. The omnibus tests yield (9) is far from
significant. Thus there is no statistical reason to suspect
a difference in the frequencies at this locus. The test for
fit based on D, using equation (6), shows surprisingly small
values of chi-square in all cases. This is mainly due to
the fact that $n_{oo}=0$ and thus $\hat{r}=0$ so that this asymptotic
approximation may not be valid. The second locus yields
significant individual chi-squares for both B27 and SIN2.
Since SIN2 was suspected as an antigen possibly associated
with NPC before the study results were known, it is
appropriate to use its associated P-value without
correction. However, B27, after multiplying its P-value by
15 (the number of remaining alleles), is clearly not
significant. The omnibus tests yield, $\chi^2_{A.}$, is also far from
significant.

The B-locus illustrates the somewhat anomalous result
that can occur from applying several individual tests and a
single multivariate test to the same data. Here we have the
situation that two individual tests are each individually
significant while the multivariate or omnibus test is not.
The most extreme such case is Rao's (1966) paradox. This is
the situation where two individual tests are significant and
the corresponding bivariate test is not significant. Healy

(1969) has shown graphically that this results from the fact that the rejection region for the individual test is the area outside of a square while the rejection region for the bivariate test is the area outside of an overlapping ellipse. In higher dimensional space these correspond to hyper-cubes and hyper-ellipsoids. Thus there are alternative hypotheses for which the individual tests (with or without adjusted P-values) are more powerful than the omnibus test and other hypotheses for which the omnibus test is more powerful. The reader is referred to Healy's lucid discussion of this tricky question.

## III.  STATISTICAL ANALYSES OF PHENOTYPIC DATA

The analyses based on the antigen frequencies above do not exploit all the information gained from the H-W model. This is clearly seen from the fact that Bernstein's simple estimator of a particular gene frequency is completely independent of the result of other antigen typings. Thus the estimator is the same whether the group is typed for only the antigen in question or for several other antigens as well. More information is available in the complete array of data (see Table 1). Thus Bernstein's simple estimator is fully efficient only when a single antigen is typed. We consider in this section analyses which will be more efficient when the phenotypic data are available.

A. Gene Frequency Estimation and Fitting the H-W Model

The MLE's, $\tilde{p}_i$, are fully efficient estimators, for the gene
frequencies. The log-likelihood is:

$$\ell n L(\underline{p}) = \Sigma_i n_i \ell n(p_i{}^2 + 2p_i r) + \underset{i<j}{\Sigma \Sigma} \; n_{ij} \ell n \; (2p_i p_j) + 2n_{oo} \; \ell n r.$$

The likelihood equations are:

$$\frac{\partial \ell n L(\tilde{\underline{p}})}{\partial p_i} = \frac{G_i}{\tilde{p}_i} + \frac{n_i}{\tilde{p}_i + 2\tilde{r}} - \frac{2n_{oo}}{\tilde{r}} - 2\Sigma \frac{n_j}{\tilde{p}_j + 2\tilde{r}} = 0, \quad i = 1, 2, \ldots, m-1, \tag{10}$$

where $\tilde{r} = 1 - \Sigma_i \tilde{p}_i$. These can be solved by the scoring
method, (see e.g., Rao, 1965, p. 302ff.) for which we need
the information matrix. This is given by

$$\underline{I}_p(\underline{p}) = \left[ E \left( - \frac{\partial^2 \ell n L}{\partial p_i \partial p_j} \right) \right], \qquad i,j = 1, \ldots, m-1, \tag{11}$$

where

$$E \left( - \frac{\partial^2 \ell n L}{\partial p_i{}^2} \right) = N \left( 1 + \frac{2}{p_i} - \frac{3p_i}{p_i + 2r} + 4 \underset{k}{\Sigma} \frac{p_k}{p_k + 2r} \right)$$

and

$$E \left( - \frac{\partial^2 \ell n L}{\partial p_i \partial p_j} \right) = 2N \left( 1 - \frac{p_i}{p_i + 2r} - \frac{p_j}{p_j + 2r} + 2 \underset{k}{\Sigma} \frac{p_k}{p_k + 2r} \right).$$

The iterative procedure starts with an initial set of
estimators, $\tilde{\underline{p}}^{(0)}$, such as the simple Bernstein's estimators
(2). The first corrected set of estimators, $\tilde{\underline{p}}^{(1)}$, is found
from the following formulas:

$$\tilde{\underline{p}}^{(1)} = \tilde{\underline{p}}^{(0)} + [\underline{I}_p(\tilde{\underline{p}}^{(0)})]^{-1} \underline{S}(\tilde{\underline{p}}^{(0)}) \tag{12}$$

and

$$\tilde{r}^{(1)} = 1 - \Sigma \tilde{p}_i{}^{(1)},$$

where $\underline{S}(\tilde{\underline{p}}^{(0)}) = [\partial \ell n \; L(\underline{p})/\partial p_1, \ldots, \partial \ell n \; L(\underline{p})/\partial p_{m-1}]$ evaluated at $p = \tilde{p}^{(0)}$. This procedure, which involves a matrix inversion at each step, is repeated until convergence is obtained. The final value of $[\underline{I}_p(\tilde{p})]^{-1}$ is the variance - covariance matrix of the MLE's. Note, of course, the estimated $\tilde{p}$'s are constrained to be non-negative and $\Sigma \tilde{p}_i \leq 1$. If $n_{oo} = 0$, the likelihood is maximized on the boundary, that is, $\tilde{r} = 0$, and $\tilde{p}_i = (G_i + n_i)/(2N)$, $i = 1, 2, \ldots, m-1$, and $\Sigma_i \tilde{p}_i = 1$.

It is possible to obtain efficient estimates without inverting the information matrix. The counting method of Smith (Ceppellini, Siniscalco, and Smith, 1955, Smith, 1957, 1967) can be done on an ordinary desk calculator. This iterative procedure is

$$\tilde{p}_i^{(1)} = \frac{G_i}{2N} + \frac{\tilde{p}_i^{(0)}}{\tilde{p}_i^{(0)} + 2\tilde{r}^{(0)}} \quad \frac{n_i}{2N} , \quad i = 1, 2, \ldots, m-1, \quad (13)$$

where $\tilde{r}^{(1)} = 1 - \Sigma_i \tilde{p}_i^{(1)}$. The estimates obtained at covergence are equivalent to the MLE's in asymptotic properties and their estimated variances may be obtained either from the inverse of the information matrix evaluated at the estimates or by an approximate method given by Smith (1967).

If the modified Bernstein's estimator is used as the initial estimator in (13), Nam and Gart (1976) show that the solution to the first iterate, $\tilde{p}_i^{(1)}$, $i = 1, 2, \ldots, m-1$, will always have asymptotic variances better than either the simple or modified Bernstein's method when $m \geq 4$.

In all of the above we assume that any antigen, $A_i$, for which $G_i = 0$, is omitted from the calculations and the magnitude of m is reduced accordingly.

The goodness of fit of the H-W model can be tested either by the usual Pearson chi-square or the likelihood ratio test of a multinomial distribution. It should be noted that the number of independent variates is $\{m(m-1)\}/2$ and m-1 independent parameters are estimated, so that the degrees of freedom of the chi-squares are each $\{(m-1)(m-2)\}/2$.


B.  Testing for Differences in Individual Gene Frequencies

The differences between the gene frequencies of two groups may be tested by the usual normal deviate test of the difference in the MLE's $(\tilde{p}_{i1} - \tilde{p}_{i2})$ divided by the estimated standard error $\{I_{P1}{}^{ii}(\tilde{\underline{p}}) + I_{P2}{}^{ii}(\tilde{\underline{p}})\}^{1/2}$, where the superscripts indicate the diagonal elements of the inverse of the information matrix each evaluated at the maximum of the two individual likelihoods. Alternatively the likelihoods may be pooled, yielding the subsequent likelihood equations:

$$\frac{\partial \ln L.(\tilde{\underline{p}}^*)}{\partial p_i} = \frac{\partial \ln L_1(\tilde{\underline{p}}^*)}{\partial p_i} + \frac{\partial \ln L_2(\tilde{\underline{p}}^*)}{\partial p_i} = 0, \qquad i = 1, 2, \ldots, m-1,$$

or, equivalently

$$\underline{S}_1(\tilde{\underline{p}}^*) + \underline{S}_2(\tilde{\underline{p}}^*) = 0, \qquad (14)$$

with the estimated pooled information matrix:

$$I_{P.}(\tilde{\underline{p}}^*) = I_{P1}(\tilde{\underline{p}}^*) + I_{P2}(\tilde{\underline{p}}^*).$$

The appropriate normal deviate test under the "total" null hypothesis: $p_{i1} = p_{i2}$, $i=1,\ldots,m-1$, is then

$$\chi^2_{Pi} = (z_{Pi})^2 = (\tilde{p}_{i1} - \tilde{p}_{i2})^2 / \{I_{P.}^{\;ii}(\tilde{\underline{p}}^*)N.(1/N_1 + 1/N_2)\}, \qquad (15)$$

where $I_{P.}^{\;ii}(\tilde{\underline{p}}^*)$ is the diagonal element of the pooled information matrix evaluated at the maximum of the pooled likelihood, $\underline{p} = \tilde{\underline{p}}^*$. An even more "correct" test would estimate $I^{ii}$'s under the hypothesis $p_{i1} = p_{i2}$ while allowing $p_{j1} \neq p_{j2}$, for $j \neq i$, in maximizing the likelihood.

As with individual tests of antigen frequencies it may be necessary to adjust the observed P-value for multiple testing.


C.  Omnibus Tests of Gene Frequency Differences

One test of the possible overall difference in two groups could be done by arranging the phenotypic data linearly in two rows yielding a $2 \times (m^2-m+2)/2$ contingency table. The usual chi-square test with $m(m-1)/2$ degrees of freedom could then be performed. However, as Rao (1965, pp. 333-334) points out, this test is not the best available if the H-W model holds for the populations being compared. Then the intrinsic difference, if any, in the two groups is between their underlying gene frequencies.

Under the H-W model, there are three omnibus tests of the hypothesis $H_o$: $p_{i1} = p_{i2}$, $i=1,2,\ldots,m-1$, which may be applied to the comparison of two sets of phenotypic data. The first is the likelihood ratio (LR) test. In this case the chi-square with m-1 degrees of freedom is:

$$\chi^2_{LR} = 2 \{\ell n \ L_1 \ (\tilde{\underline{p}}_1) + \ell n \ L_2 \ (\tilde{\underline{p}}_2) - \ell n \ L. \ (\tilde{\underline{p}}^*)\},$$

where $\tilde{\underline{p}}_1$ and $\tilde{\underline{p}}_2$ are the MLE's of the gene frequencies for
the first and second groups respectively under the
alternative hypothesis. As before, any antigen which is
found in neither cases nor controls is excluded from the
calculations of the likelihoods and degrees of freedom. An
approximation to the likelihood ratio test may be based on
Pearson's chi-square applied to the fitted values under the
null and alternative hypothesis. A third test, also with
m-1 degrees of freedom, is Fisher's discrepancy chi-square
(see Rao, 1952, pp. 186-187). This has also been employed
by Sharan (1970) in comparing ABO blood group frequencies.
The chi-square statistic is the sum of two quadratic forms:

$$\chi^2_{FP} = \underline{S}'_1(\tilde{\underline{p}}^*)[\underline{I}_{P1}(\tilde{\underline{p}}^*)]^{-1}\underline{S}_1(\tilde{\underline{p}}^*) + \underline{S}'_2(\tilde{\underline{p}}^*)[\underline{I}_{P2}(\tilde{\underline{p}}^*)]^{-1}\underline{S}_2(\tilde{\underline{p}}^*).$$

Recall from (14) that $\underline{S}_1(\tilde{\underline{p}}^*) = - \ \underline{S}_2(\tilde{\underline{p}}^*)$. This dependence in
the quadratic forms implies that $\chi^2_{FP}$ also has m-1 degrees of
freedom. As $\chi^2_{FP}$ is a special case of Rao's test for the
composite hypothesis (Rao, 1965, p. 350-352) it has the same
asymptotic distribution as the likelihood ratio test. It
may be conjectured that this test, like the corresponding
test for a single parameter, may be locally more powerful
than the likelihood ratio test.

D.  Comparison of Phenotypic Data in Two Control Groups

Rogentine, Yankee, Gart, Nam, and Trapani (1972) typed 200
controls for HLA. Sometime later a second set of 201

controls was similarly typed. The two sets of phenotypic
data are given for the A locus (m=9) in Table 4. Before
pooling the data, it was of interest to test statistically
whether the gene frequencies differed significantly in any
respect.

The analyses of the data are given in Table 5. Both
groups fit the H-W model quite well when tested by either
the Pearson's chi-square or the likelihood ratio test.
None of the individual tests are significant. The omnibus
tests yield $\chi^2_{FP}$ = 10.17 and $\chi^2_{LR}$ = 10.18, each with 8 degrees
of freedom. It is interesting to note that the omnibus test
of antigen frequencies yields a quite similar result,
$\chi^2_{A.}$ = 10.50. This reflects the fact that Bernstein's simple
method and thus the antigen frequency analyses have high
relative efficiency for this profile of gene frequencies
(Nam and Gart (1976)). At the B-locus (not given here), the
various statistical tests show no significant differences
between the two groups.


IV.  THE COMPARISON OF GENOTYPIC AND PHENOTYPIC DATA

Sometimes in case-control studies, it is possible, through
typing of the family members of the cases, to establish the
genotype of each case. That is, one can decide if phenotype
$A_i$ is actually an $A_iA_i$ or an $A_i0$. The notation for such
data has been given in Table 1. The statistical analysis of
the gene frequencies is then quite simple. We consider here

Table 4    Two Control Groups:  Phenotypic Data, LA series (First locus)
           Control group I ($N_1$=200)

| Antigens | A1 | A2 | A3 | A9 | A10 | A11 | W19(Th.) | W28(Lc17) | 0 | $G_i$ |
|---|---|---|---|---|---|---|---|---|---|---|
| A1 | 16 | 26 | 13 | 7 | 2 | 4 | 2 | 0 | – | 70 |
| A2 | 26 | 25 | 6 | 15 | 3 | 3 | 3 | 2 | – | 83 |
| A3 | 13 | 6 | 9 | 6 | 3 | 3 | 2 | 1 | – | 43 |
| A9 | 7 | 15 | 6 | 12 | 2 | 5 | 4 | 2 | – | 53 |
| A10 | 2 | 3 | 3 | 2 | 4 | 2 | 2 | 1 | – | 19 |
| A11 | 4 | 3 | 2 | 5 | 2 | 0 | 0 | 0 | – | 17 |
| W19(Th.) | 2 | 3 | 1 | 4 | 2 | 0 | 7 | 1 | – | 21 |
| W28(Lc17) | 0 | 2 | 1 | 2 | 1 | 0 | 1 | 2 | – | 9 |
| 0 | – | – | – | – | – | – | – | – | 5 | – |

Table 4    (continued)
Control group II ($N_2=201$)

| Antigens | A1 | A2 | A3 | A9 | A10 | A11 | W19(Th.) | W28(Lc17) | 0 | $G_i$ |
|---|---|---|---|---|---|---|---|---|---|---|
| A1 | 11 | 19 | 8 | 3 | 3 | 4 | 4 | 2 | – | 54 |
| A2 | 19 | 26 | 10 | 16 | 6 | 4 | 7 | 6 | – | 94 |
| A3 | 8 | 10 | 13 | 3 | 9 | 3 | 2 | 2 | – | 50 |
| A9 | 3 | 16 | 3 | 6 | 3 | 3 | 2 | 2 | – | 38 |
| A10 | 3 | 6 | 3 | 3 | 0 | 1 | 2 | 1 | – | 25 |
| A11 | 4 | 4 | 3 | 3 | 1 | 3 | 0 | 0 | – | 18 |
| W19(Th.) | 4 | 7 | 2 | 2 | 2 | 0 | 4 | 2 | – | 23 |
| W28(Lc17) | 2 | 6 | 2 | 2 | 1 | 0 | 2 | 2 | – | 17 |
| 0 | – | – | – | – | – | – | – | – | 9 | – |

Table 5   Comparison of Gene Frequencies in Control Groups Using Phenotypic Data, MLE, and Discrepancy and LR Chi-squares

| Antigens | $\tilde{p}_i \pm SE(\tilde{p}_i)$ Control I | $\tilde{p}_i \pm SE(\tilde{p}_i)$ Control II | Equation (15) $(\tilde{p}_{i1}-\tilde{p}_{i2}) \pm SE$ | Equation (15) $\chi^2_{Pi}$ |
|---|---|---|---|---|
| HLA-A1 | 0.1906 ± 0.0205 | 0.1431 ± 0.0180 | 0.0475 ± 0.0273 | 3.02 |
| -A2 | 0.2351 ± 0.0223 | 0.2640 ± 0.0233 | -0.0253 ± 0.0323 | 0.61 |
| -A3 | 0.1137 ± 0.0163 | 0.1344 ± 0.0175 | -0.0207 ± 0.0239 | 0.75 |
| -A9 | 0.1422 ± 0.0180 | 0.0982 ± 0.0152 | 0.0440 ± 0.0235 | 3.51 |
| -A10 | 0.0489 ± 0.0109 | 0.0622 ± 0.0122 | -0.0133 ± 0.0164 | 0.66 |
| -A11 | 0.0425 ± 0.0102 | 0.0458 ± 0.0105 | -0.0033 ± 0.0147 | 0.05 |
| W19(Th.) | 0.0552 ± 0.0116 | 0.0588 ± 0.0119 | -0.0036 ± 0.0166 | 0.05 |
| W28(Lc17) | 0.0229 ± 0.0075 | 0.0429 ± 0.0102 | -0.0200 ± 0.0127 | 2.48 |
| 0 | 0.1489 ± 0.0217 | 0.1506 ± 0.0216 | -0.0017 ± 0.0306 | 0.00 |
| Pearson's test of fit of H-W Law | 30.69(28d.f.) P ≈ 0.35 | 33.92(28d.f.) P ≈ 0.20 | Omnibus tests (each 8 d.f.) | |
| Likelihood ratio test of fit of H-W Law | 34.99(28d.f.) P ≈ 0.17 | 36.13(28d.f.) P ≈ 0.14 | $\chi^2_{FP}=10.17(P\approx0.24)$ $\chi^2_{LR}=10.18(P\approx0.25)$ | |

this estimation as well as the comparison of groups with genotypic data to those with phenotypic data.

A.   Gene Frequency Estimation and Fitting the H-W Model for Genotypic Data

When genotypic data are available, the likelihood equations are:

$$\frac{\partial \ell n\ L_G\ (\tilde{\underline{p}})}{\partial p_i} = \frac{n_{ii} + G_i}{\tilde{p}_i} - \frac{n_{oo} + n_{o.}}{\tilde{r}} = 0, \quad i=1,2,\ldots,m-1.$$

These can be easily solved to yield the explicit solutions:

$$\tilde{p}_i = (n_{ii} + G_i)/(2N), \quad i=1,2,\ldots,m-1,$$

and

$$\tilde{r} = (n_{oo} + n_{o.})/(2N).$$

The information matrix is $\underline{I}_G\ (\underline{p})$, where

$$I_{ii}(\underline{p}) = 2N \left( \frac{1}{p_i} + \frac{1}{r} \right), \quad i=1,2,\ldots,m-1,$$

and

$$I_{ij}(\underline{p}) = 2N/r, \quad i \neq j = 1,2,\ldots,m-1.$$

This matrix is easily inverted in general, and the exact variances and covariances of these estimators are found to be

$$V_G(\tilde{p}_i) = \{p_i(1-p_i)\}/(2N)$$

and

$$C_G(\tilde{p}_i, \tilde{p}_j) = - p_i p_j/(2N).$$

Tests of fit may be based on the Pearson's chi-square or the likelihood ratio tests. Here the degrees of freedom are $m(m+1)/2$.

B.  Testing for Differences in Individual Gene Frequencies

The analyses of Section III can be extended in a
straightforward fashion to comparing gene frequencies, one
set of which is estimated from genotypic data (cases) and
the other from phenotypic (controls).  One normal deviate
test is the difference in the MLE's, $(\tilde{p}_{i1} - \tilde{p}_{i2})$, divided by
its estimated standard error $[\{\tilde{p}_i(1-\tilde{p}_i)/(2N_1)\}+I_p^{ii}(\tilde{\underline{p}})]^{1/2}$,
where $I_p^{ii}(\tilde{\underline{p}})$ is the appropriate diagonal element of the
inverse of $I_p(\tilde{\underline{p}})$, the information matrix for phenotypic
data.

Alternatively the individual tests may be performed
using variances calculated under the null hypothesis that
all m corresponding gene frequencies in the two groups are
equal.  The log-likelihoods are added with the resulting
likelihood equation:

$$\frac{\partial \ell n\ L_.\ (\tilde{\underline{p}}^*)}{\partial p_i} = \frac{\partial \ell n\ L_1\ (\tilde{\underline{p}}^*)}{\partial p_i}+\frac{\partial \ell n\ L_2\ (\tilde{\underline{p}}^*)}{\partial p_i}=0,\ i=1,2,\ldots,m-1.$$

The resulting individual tests are:

$$\chi^2_{PGi}=(z_{PGi})^2=\frac{(\tilde{p}_{i1}-\tilde{p}_{i2})^2}{[\{\tilde{p}_i^*(1-\tilde{p}_i^*)/(2N_1)\}+I_p^{ii}(\tilde{\underline{p}}^*)]},\qquad i=1,\ldots,m-1. \qquad (16)$$

As before it may be necessary to adjust the resulting
observed P-value for multiple testing.

C.  Omnibus Tests of Gene Frequency Differences

As before we may compare all the gene frequencies in the two
groups by using Fisher's discrepancy chi-square.  In this
case this is written:

$$\chi^2_{FPG} = \underline{S}_1{'}(\tilde{\hat{p}}^*)[\underline{I}_G(\tilde{\hat{p}}^*)]^{-1}\underline{S}_1(\tilde{\hat{p}}^*) + \underline{S}_2{'}(\tilde{\hat{p}}^*)[\underline{I}_P(\tilde{\hat{p}}^*)]^{-1}\underline{S}_2(\tilde{\hat{p}}^*),$$

which has m-1 degrees of freedom.  Alternatively one may

test this hypothesis by a  likelihood ratio test.


D.  Comparison of Genotypic and Phenotypic Data:  ALL
    Patients *vs*. a Control Group

In a study of ALL patients by Rogentine *et al*. (1972),

genotypic data were available on the group of 50 cases.   The

data are given in Table 6.   These were compared to a set of

200 controls (Control Group I given in Table 4).   The data

are analyzed in Table 7.   It is seen that the disease group

is fit very well by the H-W law.   The individual chi-squares


Table 6     ALL Patients:  Genotypic Data
            (Source:  Rogentine *et al*. (1972))

Locus A (LA series), Leukemic patients ($N_2$=50)

| HLA- | Leukemic patients ($N_2$=50) | | | | | | | | | $G_i$ |
|---|---|---|---|---|---|---|---|---|---|---|
|  | A1 | A2 | A3 | A9 | A10 | A11 | W19(Th.) | W28(Lc17) | 0 |  |
| A1 | 0 | 6 | 4 | 2 | 0 | 1 | 0 | 0 | 3 | 16 |
| A2 | 6 | 6 | 9 | 3 | 1 | 3 | 6 | 1 | 1 | 36 |
| A3 | 4 | 9 | 0 | 1 | 0 | 0 | 1 | 0 | 0 | 15 |
| A9 | 2 | 3 | 1 | 0 | 0 | 1 | 0 | 0 | 0 | 7 |
| A10 | 0 | 1 | 0 | 0 | 0 | 1 | 0 | 0 | 0 | 2 |
| A11 | 1 | 3 | 0 | 0 | 1 | 0 | 0 | 0 | 0 | 5 |
| W19(Th.) | 0 | 6 | 1 | 1 | 0 | 0 | 0 | 0 | 0 | 8 |
| W28(Lc17) | 0 | 1 | 0 | 0 | 0 | 0 | 0 | 0 | 0 | 1 |
| 0 | 3 | 1 | 0 | 0 | 0 | 0 | 0 | 0 | $n_{oo}$=0 | $n_{o.}$ =4 |

Table 7    Analysis of Case-Control Study of ALL and HLA (Rogentine $et$ $al.$ (1972), First or LA Locus)

| | Gene frequencies (MLE's) | | Equation (16) | |
| | Cases ($N_1$=50) | Control ($N_2$=200) | $(\tilde{p}_{i1}-\tilde{p}_{i2})\pm SE$ | $\chi^2_{PGi}$ |
|---|---|---|---|---|
| HLA-A1 | 0.1600 ± 0.0367 | 0.1906 ± 0.0205 | -0.0306 ± 0.0438 | 0.49 |
| A2 | 0.4200 ± 0.0494 | 0.2351 ± 0.0223 | 0.1849 ± 0.0506 | 13.35[1] |
| A3 | 0.1500 ± 0.0357 | 0.1137 ± 0.0163 | 0.0363 ± 0.0368 | 0.97 |
| A9 | 0.0700 ± 0.0255 | 0.1422 ± 0.0180 | -0.0722 ± 0.0376 | 3.77 |
| A10 | 0.0200 ± 0.0140 | 0.0489 ± 0.0109 | -0.0289 ± 0.0228 | 1.61 |
| A11 | 0.0500 ± 0.0218 | 0.0425 ± 0.0102 | 0.0075 ± 0.0230 | 0.11 |
| W19(Th.) | 0.0800 ± 0.0271 | 0.0552 ± 0.0116 | 0.0328 ± 0.0268 | 1.50 |
| W28(Lc17) | 0.0100 ± 0.0099 | 0.0229 ± 0.0075 | -0.0129 ± 0.0158 | 0.67 |
| 0 | 0.0400 ± 0.0196 | 0.1489 ± 0.0217 | -0.1089 ± 0.0383 | 8.08[2] |
| Pearson's test of fit of H-W Law | 33.65(36d.f.) P≈0.60 | 30.69(28d.f.) P≈0.35 | Omnibus tests (each 8 d.f.) | |
| Likelihood ratio test of fit of H-W Law | 32.12(36d.f.) P≈0.65 | 34.99(28d.f.) P≈0.17 | $\chi^2_{FPG}$=24.59(P≈0.002) $\chi^2_{LR}$=30.60(P≈0.0002) | |

1 P-value ≈ 0.0003  (two tailed)
2 P-value ≈ 0.005   (two tailed)

yield significant differences between HL-A2 and the
recessive gene frequencies by (16). The excess of HL-A2 in
the case group is partially compensated for by a deficiency
of recessives in that group. Both differences would be
significant even after adjustment for multiple testing,
although this may not be necessary as previous studies had
implicated HL-A2 with ALL. Fisher's discrepancy chi-square
yields an $\chi^2_{FPG}$=24.59 (8d.f.) with a P ≃ 0.002 and the
corresponding likelihood ratio test yields
$\chi^2_{LR}$=30.60(P≈0.0002). Clearly the disease and control groups
differ significantly at this locus. No significant
differences are found at the B-locus.

It was conjectured that HL-A2 may confer susceptibility
to ALL. A later study (Rogentine, Trapani, Yankee, and
Henderson, 1973) appeared to associate HL-A2 with longer
survivorship (see also Gart, 1979).


V.   THE COMPARISON OF GENOTYPIC DATA

When genotypic data are available on two groups, the
individual and omnibus tests for their comparison are
straightforward. As can be seen from Section IV A, the
MLE's of the gene frequencies and their variance-covariance
matrix are obtainable in explicit form. The tests for this
case are completely analogous in structure to those for
comparing antigen frequencies (see Section II B and (9)).
Inverting the variance-covariance matrix in the quadratic

form for this situation, Smouse (1979) showed that this

omnibus test statistic can be expressed as the usual chi-

square statistic of a 2×m contingency table.


## VI.   RELATED ANALYSES OF LINKAGE DISEQUILIBRIUM AND OF THE RELATIVE RISK

In addition to testing for differences in gene frequencies

at a single locus, the subsequent logical step is the

consideration of frequencies at pairs of loci.  This has

been considered, for instance, by Simons *et al*. (1976) in

regard to the increased frequency of the joint occurrence of

HL-A2 and SIN-2 among NPC patients.  In relation to this

concept it is useful to consider the analysis of linkage

disequilibrium.  Such analyses are reviewed by Cavalli-

Sforza and Bodmer (1971, pp. 246-248 and pp. 283-288) in

relation to the HLA system.  Weir (1979) is an excellent

review of this subject with an extensive bibliography.

Extension to disequilibruim in three loci in HLA systems is

considered by Piazza (1975) and Yasuda (1978).  The

difficulties involved in detecting the third order linkage

disequilibrium between two HLA loci and a disease

susceptibility gene is discussed by Thompson and Bodmer

(1979).  A method for detecting HLA haplotype-disease

association based on population data for two disease models

(Terasaki and Mickey (1975) and Thomson and Bodmer (1977))

is investigated by Porte and McHugh (1980).  Various

standard and alternative concepts and statistics for

measuring disequilibrium at multiloci such as the HLA-A, B, and C loci are studied by Karlin and Piazza (1981).

The antigen frequencies can be used to estimate the relative risk of a rare disease in the usual way. These methods have been extensively reviewed by Gart (1970, 1971, 1979) and Svejgaard, Jersild, Nielsen and Bodmer (1974). Most of the methods suggested by Gart are implemented in the computer program of Thomas (1975).

We note also that the methods herein described in III and IV are implemented in computer program of Nam (1978a, b).

## ACKNOWLEDGMENTS

We thank Mrs. Virginia Grubar for careful typing of the manuscript and Dr. William C. Blackwelder for suggestions.

## REFERENCES

Cavalli-Sforza, L.L. and Bodmer, W.F. (1971). *The Genetics of Human Populations*. Freeman, San Francisco.

Cepellini, R., Siniscalco, M., and Smith, C.A.B. (1955). The estimation of gene frequencies in a random-mating population. *Annals of Human Genetics*, 20: 97-115.

Cotterman, C.S. (1953). Regular two-allele and three-allele phenotype systems. Part I. *American Journal of Human Genetics*, 5: 193-235.

Gart, J.J. (1970). Point and interval estimation of the common odds ratio in the combination of 2x2 tables with fixed marginals. *Biometrika*, 57: 471-475.

Gart, J.J. (1971). The comparison of proportions: A review of significance tests, confidence intervals and adjustments for stratification. *International Statistical Review*, 39: 148-169.

Gart, J.J. (1979). Statistical analyses of the relative
    risk. *Environmental Health Perspectives*, 32:
    157-167.

Healy, M.J.R. (1969). Rao's paradox concerning multivariate
    tests of significance. *Biometrics*, 25: 411-413.

Karlin, S. and Piazza, A. (1981). Statistical methods for
    assessing linkage disequilibrium at the HLA-A, B, C
    loci. *Annals of Human Genetics*, 45: 79-94.

Nam, J. (1978 a). Discrepancy chi-square method for testing
    the homogenity of two phenotype groups with respect
    to the HLA system. Unpublished manuscript.

Nam, J. (1978 b). Discrepancy chi-square method for testing
    the homogenity of two groups with respect to HLA
    system when one group is genotypic and the other is
    phenotypic. Unpublished manuscript.

Nam, J., and Gart, J.J. (1976). Bernstein's and gene-
    counting methods in generalized ABO-like systems.
    *Annals of Human Genetics*, 39: 361-373.

Noether, G.E. (1955). On a theorem of Pitman. *Annals of
    Mathematical Statistics*, 26: 64-68.

Piazza, A. (1975). Haplotypes and linkage disequilibria
    from the three-locus phenotypes. In *Histo-
    compatibility Testing 1975* (F. Kissmeyer-Nielsen,
    ed.). Munksgaard, Copenhagen, pp. 923-927.

Porte, J. and McHugh, R. (1980). Detection of HLA haplotype
    associations with disease. *Tissue Antigens*, 15:
    337-345.

Rao, C.R. (1952). *Advanced Statistical Methods in Biometric
    Research*. Wiley, New York.

Rao, C.R. (1965). *Linear Statistical Inference and Its
    Applications*. Wiley, New York.

Rao, C.R. (1966). Covariance adjustment and related
    problems in multivariate analysis. In *Multivariate
    Analysis* (P.R. Krishnaiah, ed.) Academic Press, New
    York, pp. 87-103.

Rogentine, G.N. Jr., Yankee, R.A., Gart, J.J., Nam, J., and
    Trapani, R.J. (1972). HL-A antigens and disease.
    Acute lymphocytic leukemia. *Journal of Clinical
    Investigation*, 51: 2420-2428.

Rogentine, G.N., Trapani, R.J., Yankee, R.A., and Henderson,
    E.S. (1973). HL-A antigens and acute lymphocytic

leukemia: The nature of the HL-A2 association. *Tissue Antigens*, 3: 470-476.

Ryder, L.P. and Svejgaard, A. (1976). *Association between HLA and Disease*. Report from the HLA and disease registry of Copenhagen. (Reprinted as appendix to Mourant, A.E., Kopec, A.C., and Domaniewska-Sobczak, K. (1978). *Blood Groups and Disease*. Oxford University Press, Oxford, pp. 275-298).

Sharan, J. (1970). Statistical analysis of ABO blood group data from Bihar. *Sankhyā, B, 32*: 27-30.

Simons, M.J., Wee, G.B., Goh, E.H., Chan, S.H., Shanmugaratnam, K., Day, N.E., and de-Thé, G. (1976). Immunogenetic aspects of nasopharyngeal carcinoma. IV. Increased risk in Chinese of nasopharyngeal carcinoma associated with a Chinese-related HLA profile (A2, Singapore 2). *Journal of the National Cancer Institute*, 57: 977-980.

Smith, C.A.B. (1957). Counting methods in genetical statistics. *Annals of Human Genetics*, 21: 254-276.

Smith, C.A.B. (1967). Notes on gene frequency with multiple alleles. *Annals of Human Genetics, 31*: 99-107.

Smouse, P.E. (1979). Statistical analysis of HLA-disease associations. In *Genetic Analysis of Common Diseases: Application to Preventive Factors in Coronary Disease* (C.F. Sing and M. Skolnick, eds.). Alan R. Liss, Inc., New York, pp. 545-551.

Stevens, W.L. (1950). Statistical analysis of the A-B-O blood groups. *Human Biology, 22*: 191-217.

Svejgaard, A., Jersild, C., Nielsen, L.S., and Bodmer, W.F. (1974). HL-A antigens and disease, statistical and genetical considerations. *Tissue Antigens, 4*: 95-105.

Svejgaard, A., Platz, P., Ryder, L.P., Nielsen, L.S., and Thomsen, M. (1975). HL-A and disease associations. A survey. *Transplantation Reviews, 22*: 3-43.

Terasaki, P.I. (1980). *Histocompatibility Testing 1980*. Report of the Eighth International Histocompatibility Workshop held in Los Angeles, California, U.S.A., February 4-10, 1980.

Terasaki, P.I. and Mickey, M.R. (1975). HL-A haplotypes of 32 diseases. *Transplantation Reviews, 22*: 105-119.

Thomas, D.G. (1975). Exact and asymptotic methods for the combination of 2×2 tables. *Computers and Biomedical Research*, 8: 423-446.

Thomson, G. and Bodmer, W.F. (1977). The genetic analysis of HLA and disease associations. In *HLA and Disease* (J. Dausset and A. Svejgaard, eds.). Munksgaard, Copenhagen, pp. 84-93.

Thomson, G. and Bodmer, W.F. (1979). HLA haplotype associations with disease. *Tissue Antigens*, 13: 91-102.

Weir, B.S. (1979). Inferences about linkage disequilibrium. *Biometrics*, 35: 235-254.

Wilks, S.S. (1962). *Mathematical Statistics*. Wiley, New York.

Yasuda, N. (1978). Estimation of haplotype frequency and linkage disequilibrium parameter in the HLA system. *Tissue Antigens*, 12: 315-322.

Yasuda, N. and Kimura, M. (1968). A gene-counting method of maximum likelihood for estimating gene frequencies in ABO and ABO-like systems. *Annals of Human Genetics*, 31: 409-420.

# 7

# EVALUATION OF SCREENING PROGRAMS FOR THE EARLY DETECTION OF CANCER

Philip C. Prorok
National Cancer Institute
Bethesda, Maryland

## I. INTRODUCTION

A number of national and international conferences have been
held during the past several years to discuss and evaluate
screening for the early detection of cancer. The National
Cancer Institute has conducted state-of-the-art conferences
on screening for cancers of the breast, colon, lung, bladder
and cervix. The International Union Against Cancer
sponsored a meeting on cancer screening involving
participants from several nations. This interest in
screening programs is part of an increasing emphasis on
prevention in cancer. Along with this increased emphasis
there is an increased responsibility to thoroughly evaluate

new prevention approaches before they are recommended for widescale application. This chapter will focus on the rigorous scientific evaluation of early detection programs for cancer. Many of the ideas apply to screening for other diseases as well.

In any discussion of screening, one must first be clear about the definition being used. Several definitions of screening have been offered (for example, see Wilson and Jungner (1968) and a series of articles from *The Lancet*, October 5 to December 21, 1974). The definition adopted here from Sackett and Holland (1975) is that screening is the testing of "apparently healthy volunteers from the general population for the purpose of separating them into groups with high and low probabilities for a given disorder ... the encounter is initiated by those who do the tests. However, the objective of screening is unique: the early detection of those diseases whose treatment is either easier or more effective when undertaken at an earlier point in time. There is thus an implicit promise that those who volunteer to be screened will benefit (i.e., that they will be followed up to exact diagnosis and long-term care and will receive treatments of proven efficacy)."

Sackett and Holland also stated that screening is to be contrasted with related procedures of epidemiological surveys, case finding and diagnosis. An epidemiological survey involves the determination of various demographic, social and biological characteristics of selected population

samples. The objective is to obtain new knowledge, as in a
study of disease natural history. The encounter is
initiated by those who do the survey tests, but no health
benefit to the participant is implied. Case-finding is
quite different from screening and surveys in that this
involves the application of a test or tests for certain
disorders to patients who have sought medical care for
reason of a complaint which may be unrelated to the disorder
which the test seeks to detect. The encounter is initiated
by the patient for the purpose of a health assessment.
There is no implied guarantee of benefit to the patient,
only the assurance that he will receive the best available
care. Diagnosis is the application of a variety of
examinations to patients who have sought health care in
order to pinpoint as accurately as possible the exact cause
for the patients' signs or symptoms. Diagnosis can often be
an extensive, complicated and costly procedure. It
ordinarily follows the detection of disease by one of the
other three maneuvers.

As screening is initiated by those doing the testing
and as there is an implicit promise of benefit to apparently
healthy individuals who volunteer for a screening program,
the evaluation of a screening program before it is
recommended for widescale implementation must be considered
very carefully. Far greater certainty of efficacy is
required when patients are solicited through screening than
when individuals seek medical care because of symptoms, as

discussed by Sackett and Holland (1975). Rigorous
scientific evaluation is necessary. The method of choice is
the randomized controlled trial (RCT). This stringent
approach is justified for a number of reasons summarized
Sackett (1975):  the interpretation of nonexperimental data
encounters several pitfalls which are virtually
insurmountable with existing methodology; the economic and
social costs of screening and subsequent follow-up and
treatment are too great to allow the decision of whether or
not to screen to be based upon nonexperimental evidence (and
furthermore the magnitude of both benefit and cost can be
quantified in a properly designed experimental study); if a
screening program is widely advocated on the basis of
nonexperimental data and is subsequently shown to be
ineffective, the resulting public and professional
skepticism could jeopardize future prevention programs of
proven value; the search for improved screening methods for
a given disease may be retarded by the mass implementation
of an existing test and if the existing test is based on
nonexperimental evidence and is later shown to be of no
value, valuable time will have been wasted; finally, the
feasibility of performing large scale RCTs in screening has
been established.

The discussion in the following sections of this
chapter is focused on the RCT framework for screening
evaluation.  Properties of screening tests are addressed
briefly and experimental and observational studies are

compared. The lead time and length biases inherent in
screening and their impact upon various study endpoints are
considered. Several RCT study designs are described,
followed by mention of subsidiary evaluation procedures.


II.  ISSUES IN THE EVALUATION OF SCREENING PROGRAMS

A.  Screening Test and Screening Program

It is important to distinguish between a screening test and
a screening program when evaluating early detection
procedures as each has its own set of evaluation criteria.
A screening test is some form of examination or combination
of examinations performed on individuals for the purpose of
separating them into risk categories. The examination may
be in the form of a physical exam, an Xray, a blood or urine
test, a questionnaire, and so on. A screening program
involves the performance of a screening test on a population
at certain ages or frequencies with provision for follow-up
to diagnosis and treatment for individuals with a positive
test result.

While the screening program is of primary interest in
this discussion, it is useful to consider briefly the
desirable properties of a screening test. Numerous authors
have written about this subject (see for example, Wilson and
Jungner (1968), Thorner and Remein (1961), Buck and Gart
(1966), Gart and Buck (1966), Galen and Gambino (1975), and
Lilienfeld (1974)). The test should be safe, rapid,

inexpensive, relatively easy to apply and acceptable to the
people being screened. In addition, the test should possess
high sensitivity and specificity and a good predictive
value. To clarify what is being discussed, sensitivity is
the proportion designated positive by the screening test
among all individuals who have the disease while specificity
is the proportion designated negative by the test among all
those who do not have the disease. One would certainly
desire a high test sensitivity, but this is very difficult
to estimate in the screening setting because the diagnostic
workup, which presumably gives an accurate indication of
disease status, is not ordinarily performed on individuals
with a negative screening test. Other problems encountered
in the estimation of sensitivity are discussed by Cole and
Morrison (1980). While sensitivity is important, increased
emphasis has been placed recently on specificity in the
screening context since low specificity can lead to
unmanageable numbers of false positive cases which could
overwhelm diagnostic services and result in prohibitive
costs. The predictive value of a positive test is the
proportion of individuals with a positive screening test who
are found to have the disease as a result of diagnostic
workup. This quantity is a function of sensitivity,
specificity and disease prevalence, and the
interrelationships are investigated in Cole and Morrison
(1980).

These screening test properties are of greatest utility
in assessing the feasibility of a test for screening
purposes as opposed to its use in diagnosis or monitoring
therapy.  A test has different sensitivity and specificity
requirements in the different settings; that is, for
diagnosis, high sensitivity is necessary but high
specificity is not, while for screening, reasonably high
specificity is required while sensitivity need not be
extremely high.  In any case, these properties should be
estimated as accurately as possible in well designed studies
before a test is used in a screening program.  Beyond this,
the screening test properties, in themselves, tell one
nothing about the impact which screening might have on the
disease in question.  Instead, if a screening test having
good properties is available, this can be made part of a
screening program involving therapy and follow-up.  An
evaluative screening program then focuses on the central
issue:  does early detection plus treatment alter the final
disease outcome?  The emphasis of this chapter is on how one
obtains an answer to this question.

B.  Experimental and Observational Studies

There are two basic types of studies that can be undertaken
to evaluate screening programs, experimental and
observational.  The experimental study is often termed the
randomized controlled trial and is the method of choice for

reasons discussed below. Observational studies are also
referred to as nonrandomized, quasi-experimental or sub-
experimental. They may be similar to experimental studies
in many respects, but they lack the crucial element of
randomization. This section will deal with basic
prerequisites and criteria for an evaluative screening
program and then describe some pitfalls in observational
studies.

Presupposing the existence of a good screening test,
the several principles of screening which should be
considered before a screening program is begun are
summarized by Wilson and Jungner (1968) and in Miller (1978
b). The disease should be a serious health problem, being
common in occurrence and the cause of substantial mortality
and morbidity. The target population should be clearly
defined and limited to high risk individuals, in those
situations where risk factors can be determined, so as to
yield a reasonable disease prevalence and acceptable
predictive value for the screening test. The target
population should be accessible and a reasonable level of
response to the invitation to be screened should be
expected. There should exist an agreed upon and effective
treatment for the lesions discovered by screening in the
population. Further, there should be a reasonable
expectation that individuals designated as positive by the
screening test will comply with recommendations for further
diagnostic workup, and that individuals with diagnosed

disease, as well as their physicians, will comply with
recommendations for therapy and follow-up. Sufficient
resources in terms of personnel and facilities both for
screening and for subsequent diagnosis and therapy should be
available. Agreement should be reached on a policy for
early recall of individuals with suspicious findings and on
the frequency of routine recall for those with negative
findings. Uncertainty may exist about one or more of these
principles before a screening program is begun; in fact, it
may be the purpose of an evaluation study to investigate
such uncertainty. This should be clearly indicated.
Finally, a quality control procedure to maintain the
sensitivity and specificity of the test should be
established and the program should be monitored on a regular
basis.

There are certain additional criteria worthy of
consideration prior to the start of a study and which should
also be given careful attention in reporting the study
results as discussed by Sackett (1975). These apply
particularly to an evaluation study, and in part are similar
to the principles just listed. The criteria for inclusion
of an individual in the study should be clearly established.
For example, general population samples should be
distinguished from those of industrial or hospital
populations. Most importantly, it should be stated as to
whether the study individuals are asymptomatic or
symptomatic, and the criteria for this distinction should be

clear. For a strict evaluation study, the allocation of
individuals to a screened or unscreened group should be done
by randomization. A stratified randomization may be used in
which the stratification variables would be those which
separate the population into groups with differing risks of
developing and dying from the disease. The exact nature of
the screening test, diagnostic workup and therapy should be
agreed upon and described in detail. This would include a
statement about who performs and interprets the screening
test, and the sequence of steps and procedures in the
diagnostic workup that follows a positive screen. The level
of compliance, as mentioned above, with screening, diagnosis
and treatment should be reported. The problems of co-
intervention and contamination must be dealt with. Co-
intervention refers to the performance of additional
screening or other procedures beyond the one under study on
the experimental group but not on the comparison or control
group. This should be avoided. Alternatively, the extra
procedures could be performed equally on both groups. It is
particularly important that follow-up be performed with
equal vigor in both groups. Contamination refers to the
situation where members of the control group get the
screening test. To optimize the possibility of
demonstrating the effectiveness of a screening test, it
should be applied to all members of a study group and to no
members of a control group, if possible. The extent of any
contamination should be documented.

Finally, there are two important and related criteria concerned with diagnosis and mortality as discussed in Miller (1978 a, 1978 b). Newer screening modalities which reach farther and farther back into the natural history of disease detect certain entities which may or may not be disease. These have been termed borderline lesions or abnormalities. Ideally, their natural history should be understood before screening is begun, but usually this will not be the case. Nevertheless, based on available knowledge, an agreed upon policy for the classification, management and follow-up of borderline lesions should be established. The classification rules for assigning cause of death are similarly important. For reasons discussed in the next section, the key and definitive endpoint for determining screening effectiveness is mortality. The criteria for allocating deaths to various causes, particularly the disease of interest, should be specified and described in detail. Furthermore, the follow-up and reporting of mortality experience in study and control groups should include total mortality, that is, deaths from all causes, to aid in reducing bias from incorrect cause of death assignment.

Virtually all of the principles and criteria just outlined for an ideal screening evaluation study apply to both the experimental and the observational study. The key difference is randomization. That is, only the experimental study has a control group which is constructed by a chance

procedure. Lacking this, the observational study contains
certain noncomparability deficiencies which can render the
inferences drawn from such a study questionable or invalid.
Indeed, it is relatively easy to observe the outcome of a
treatment or procedure in the absence of a comparison group
and to construct hypotheses concerning the outcome.
However, such observations by themselves are poor quality
evidence because of the clear possibility of observer bias
and the absence of knowledge of what would have happened if
no treatment had been given. An improvement is the use of a
comparison group, but these can be a very mixed lot,
including such useless entities as "those who refused
treatment". What is required for rigorous evaluation is
controlled experimentation, in which the comparison or
control group is the same in all respects which might
influence the course of the disease as the study group, save
for the special procedure under study, in this case a
screening test. This is accomplished by means of the RCT.

The purposes and advantages of randomization have been
addressed by several authors (for example see Byar $et$
$al.$ (1976)). Use of a control group chosen by any method
other than randomization requires the assumption either that
the control and treatment groups are identical in all
important variables except the treatment under study or that
one can correct for all relevant differences. In the latter
case one must further assume that all factors affecting the
course of the disease are known. These assumptions are

rarely if ever justified. One is thus led to consider some
of the problems and invalid inferences that can arise in
non-RCT screening trials.

There are numerous examples documented in the
literature where use of non-randomized, usually historical,
controls has led to the wrong conclusion about the efficacy
of therapy (for example see Byar *et al.* (1976), Peto *et
al.* (1976, 1977), Scheiderman (1966), and Herbert (1977)).
In the screening setting, the ability to make valid
inferences from studies of individual patients or
observations of changes in community rates is severely
limited by the difficulty of defining appropriate comparison
groups, by the lack of detailed knowledge of disease natural
history, and by the inadequate understanding of reasons for
changes over time in incidence, survival and mortality rates
for various diseases. For example, the steadily falling
incidence and mortality rates of cancer of the stomach and
cervix in most of North America make it virtually impossible
to assess the impact of control measures for either disease
without a rigorously designed evaluation study which
includes valid (randomized) comparison groups observed for
the necessary number of years (see Henderson (1976)).

A number of suggestions have been offered, apart from a
screening effect, to explain the decline in cervical cancer
rates: antimicrobial agents which were introduced following
World War II, improvements in hygiene and medical care, and
an increasing rate of hysterectomy. The latter is directly

relevant to the problems inherent in the comparison of observational data over time. In this type of comparison, it is necessary to ensure, if possible, that the populations at risk; i.e., the denominators of the rates, are at comparable risk over the observation period. If the frequency of hysterectomy is greater for more recent time periods, this could be responsible for a bias in the trend in cervical cancer mortality as discussed by Apostolides and Henderson (1977).

The essence of the problem with observational studies is the noncomparability of groups or populations being compared. In screening, a strong selection bias operates with regard to the characteristics of individuals who volunteer to participate in a screening program compared to those who refuse to participate. This has been vividly demonstrated in the HIP breast cancer screening study.

The HIP study is a RCT designed to evaluate screening with clinical examination plus mammography by comparing breast cancer mortality between a control group and a total study group, the latter including those who were screened and those who refused screening. Analysis of the data indicated substantial differences in disease characteristics between the respondents and the refusers. The prevalence rate at the first screen was 2.73 per 1000. Among women who appeared for rescreening, the detection rate was 1.49 per 1000 person-years, while the rate among screened women whose disease was not detected by screening was 0.92 per 1000

person-years.  Overall, among study screened women over a 5
year period, the rate of case detection was about 2.2 per
1000.  By contrast, the incidence rate among study women who
refused screening was 1.45 per 1000 per year.  The rate in
the control group was 1.87 per 1000 per year.  The
relatively low rate among the refusers suggests that women
with a higher risk of breast cancer tended to select
themselves for screening.  In addition, if one examines the
death rates from all causes excluding breast cancer, the
rates per 10,000 per year are:  control group 57.6; total
study group 56.9; study screened group, 42.4; study refused
screening group; 85.6.  Thus general mortality excluding
breast cancer was far lower among the respondents (see
Shapiro (1977)).

Given this selection behavior, it would not have been
possible to construct an equivalent comparison group for the
women who elected to be screened from among the remainder of
female HIP members.  The consequence of ignoring this
selection factor and resting the case of screening benefit
in the HIP study on a comparison between the screened women
and a cross-section of other HIP members of similar ages
(e.g., the control group) in their breast cancer mortality
would have been to cast doubt on the value of screening.
This is contrary to the HIP study's conclusion that about a
30 percent reduction in breast cancer mortality in the total
study group (screening participants and refusers) is
attributable to screening (see Shapiro *et al.* (1982)).  The

erroneous conclusion might follow because while it is true
that the screened women have a lower breast cancer death
rate than the control group, they also have a lower general
mortality rate (excluding breast cancer), so that the lower
breast cancer rate could simply reflect the generally lower
mortality. An explanation is that the control group
contains two subgroups: the counterpart of the study group
women who were screened, a subgroup at low risk for general
mortality excluding breast cancer but at high risk for
breast cancer incidence and mortality, and the counterpart
of the study group women who refused screening, a subgroup
at high risk for general mortality excluding breast cancer
but at low risk for breast cancer incidence and mortality.
However, as discussed by Beahrs et al. (1979), there is no
way to subclassify the control group to achieve initial
equivalence with these two subgroups of the study group.

C.   Endpoints in Evaluation Studies

In the evaluation of any procedure, a decision must be made
as to the outcome measure which will be used as the basis
for evaluation. The key criterion or outcome measure in
screening for the early detection and treatment of cancer is
the clinical outcome of the patients identified with
disease. This is equivalent to focusing on population
mortality as the crucial endpoint (see Sackett (1975), Cole
and Morrison (1980), Beahrs et al. (1979), and Thomas et
al. (1977)). Mortality rate as used here is defined as the

number of deaths per unit time per unit population at risk.
Only by unequivocal demonstration of reduction in disease
mortality can one hope to justify the cost of screening and
fulfill the implied promise of benefit to apparently healthy
individuals who volunteer to participate in a screening
program. Drawbacks associated with other endpoints are
described in this section. These drawbacks are closely
associated with the lead time and length biases of
screening programs. These biases are also discussed.

## 1. Lead Time and Length Bias

The concepts of lead time and length bias are most
easily understood within the framework of a conceptual model
of disease and screening. The idealized disease natural
history model adopted here is the three state progressive
disease model wherein an individual in a screened population
is assumed to be in one of three states: the disease free
state, $S_0$, the preclinical disease state, $S_p$, or the
clinical disease state, $S_c$. Individuals in $S_0$ are either
disease free or have disease characteristics which are not
detectable by the screening test. Individuals in $S_p$ have
preclinical disease of which they are unaware but which is
detectable by the screening test. The transition from $S_0$
into $S_p$ is assumed to take place at the first point in time
at which disease is detectable by the screening test. In
$S_c$, the disease is characterized by overt signs or symptoms
leading to diagnosis. This state begins at the transition
from $S_p$ into $S_c$ which marks the point of clinical diagnosis.

Only the transitions $S_0 \rightarrow S_p \rightarrow S_c$ are allowed in the progressive
disease model. This basic structure underlies many of the
modelling efforts that have appeared in the literature in
articles by Zelen and Feinleib (1969), Prorok (1976 a and
b), Shwartz (1978 a and b), Albert *et al.* (1978 a and b),
and Blumenson (1976 and 1977).

Of interest is the interaction of this disease process
with a screening process which is superimposed on it. For
example, consider a periodic screening program consisting of
k+1 screens performed at times $0, \Delta, \ldots, k\Delta$, where the time of
the first screen has been arbitrarily designated as zero.
In this situation, the screening examinations are sampling
points of the disease process in that individuals who are in
$S_p$ at the time they are screened are detected with disease
or sampled. (If the test is not perfect, such an individual
may be falsely classified as negative.) Individuals who are
preclinical at some point in time or during some time
interval will collectively have some distribution of
preclinical state durations or sojourn times. The screening
process therefore samples a distribution of preclinical
state sojourn times from the disease natural history process
as discussed by Prorok (1976 a).

If one examines the natural history of a given
individual in terms of this model, the concept of  lead time
becomes clear. Figure 1 displays the relationship between
preclinical state duration, the point of detection by
screening and the  lead time. If an individual participates

DURATION OF PRECLINICAL STATE

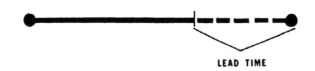

LEAD TIME

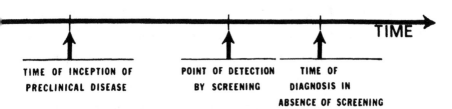

TIME

TIME OF INCEPTION OF        POINT OF DETECTION        TIME OF
PRECLINICAL DISEASE         BY SCREENING              DIAGNOSIS IN
                                                      ABSENCE OF SCREENING

Figure 1    Relationship between duration of preclinical
            disease, screening point and lead time.


in a screening program and has his disease detected earlier

than it would have been in the absence of screening, then

the amount of time by which diagnosis is advanced as a

result of screening is the  lead time, represented by the

dashed portion of the preclinical duration line.  (For

simplicity is has been assumed that detection and diagnosis

occur virtually simultaneously.)  Note that because of the

lead time phenomenon the point of diagnosis is advanced and

survival as measured from diagnosis is automatically

lengthened for cases detected by screening even if there is

no increased therapeutic benefit.  This is referred to as
lead time bias and is addressed further below.

The other screening bias, which is perhaps more subtle
but more important, is called length bias and is discussed
by Zelen (1976) and by Albert *et al.* (1978 a).  This bias is
the basis for determining which preclinical disease cases
will have a non-zero  lead time.  That is, cases of disease
detected by a screening program cannot be considered as a
representative or random sample from the general
distribution of $S_p$ sojourn times in the screened population.
Instead, the detected cases are overrepresented with longer
duration preclinical disease.  This is illustrated in Figure
2 for the case of three screens.  The sampling or detection
of preclinical cases by screening is equivalent to throwing
a grid of three dashed vertical lines at random onto the
pattern or distribution of $S_p$ sojourn times.  Detection of a
case corresponds to a vertical line intersecting a
horizontal line.  Clearly, longer horizontal lines are more
likely to intersect a vertical line than are shorter
horizontal lines; i.e., long duration preclinical disease
has a higher probability of being detected by screening than
does short duration disease.  This is length-biased sampling
or length bias.

Long duration disease can be equated with slow growing
disease, and it is likely that slower growing preclinical
disease progresses to slower growing clinical disease.  Thus
cases of disease with more favorable progression rates are

the ones most likely to be detected by screening and they
will tend to have characteristics of good prognosis, such as
lack of involvement of regional lymph nodes.  Furthermore,
these good prognosis cases would tend to have a more
favorable outcome even in the absence of screening.  As
Bailar (1976) has pointed out:  "It must be understood that
these are inherent characteristics of the disease that
increase the likelihood of cancer being detected at
screening, rather than beneficial effects of screening
resulting from detection at an earlier time.  The latter may
also happen, but it is a separate effect and cannot be shown
simply by the fact that cancers detected at screening have a
high cure rate or a low incidence of nodal involvement."

A final point about Figure 2 is in reference to the two
long horizontal lines ending in dots which are meant to
represent nonprogressive preclinical disease.  One can
modify the above progressive disease model to include a
nonprogressive preclinical disease state in which disease is
detectable by the screening test but would not progress to
clinical disease in the absence of screening before the
individual died of some other cause.  Clearly the detection
of such a case is of no benefit to the individual, but, as
the figure indicates, such cases can remain preclinical over
the course of repeated screenings and may therefore be more
likely to be detected.  Some of the borderline lesions being
found by newer screening modalities may be of this type and
their existence adds further difficulties to the evaluation

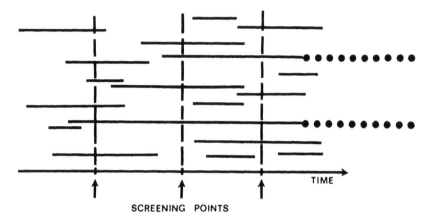

Figure 2    Length-biased sampling in repeated screening.
            Solid horizontal lines represent a distribution
            of preclinical durations.  Lines ending in dots
            represent nonprogressive preclinical disease.

of a screening program.  The magnitude of the problem of
nonprogressive disease can only be determined, however, by
the acquisition of more knowledge about the preclinical
natural history of disease.

Further insight into the length bias problem can be
gained by examining some numerical examples under the
progressive disease model.  The results which follow are for
equally spaced screens which are $\Delta$ units apart.  Individuals
in the screened population who enter $S_p$ sometime during the
screening program are first classified into generations
according to the interscreening interval during which they
become preclinical.  Thus zero[th] generation individuals are
those who are in $S_p$ at time 0, while $i$[th] generation
individuals are those who enter $S_p$ during the interval
$((i-1)\Delta, i\Delta]$, $i=1,2,\ldots,k$.  The probability density function

(p.d.f.) of preclinical sojourn times for $i^{th}$ generation
individuals in the absence of screening is denoted by $q_i(t)$
It is this set of distributions which is sampled by the
screening program.

In the usual situation when the screening test is not
perfect and so yields false negative results, one can
inquire about individuals from each generation who might be
detected at any screen which they attend after entering $S_p$.
Let $g_{ij}(t)$ be the p.d.f. of the $S_p$ sojourn time of $i^{th}$
generation individuals whose disease is potentially
detectable at the $(j+1)^{st}$ screen at time $j\Delta$. Under the
assumptions of a stationary disease process and a uniform
distribution of entry times into $S_p$ over the interval
$((i-1)\Delta, i\Delta]$ for $i^{th}$ generation individuals, it can be shown
(see Prorok (1976 a)) that

$$g_{0j}(t) = \begin{array}{l} K_{0j}(t-j\Delta) q_0(t), \quad j\Delta \leq t < \infty \\ 0, \text{ elsewhere}, \quad j=0,1,\ldots,k \end{array}$$

and

$$g_{ij}(t) = \begin{array}{l} K_{ij} \Delta^{-1} \{t-(j-i)\Delta\} q_i(t), \quad (j-i)\Delta \leq t < (j-i+1)\Delta \\ K_{ij} q_i(t), \quad (j-i+1)\Delta \leq t < \infty \\ 0, \text{ elsewhere}, \quad i=1,2,\ldots,k, \; j=i, \; i+1,\ldots,k, \end{array}$$

where $K_{0j}$ and $K_{ij}$ are normalizing constants. The $g_{ij}(t)$
are the $S_p$ sojourn time distributions of the length biased
cases, as is clear from the functional form in which the g
distribution is proportional to the product of t, the
length, and the corresponding q distribution. Larger t
values have a higher probability of being sampled.

With this result, it is of interest to determine the properties of the length biased cases at each screen in the screening program, termed local properties. These properties can be calculated from the $g_{ij}(t)$ functions using appropriate weighted averages. For example, the local $S_p$ sojourn time p.d.f. of all cases detected by screening at time $j\Delta$ is given by

$$g_{Dj}(t) = \sum_{i=0}^{j} D_{ij}\, g_{ij}(t) \; / \; (\sum_{i=0}^{j} D_{ij}), \; j=0,1,\dots,k.$$

This is a weighted combination of the p.d.f.s of cases from each generation that might be detected at time $j\Delta$. Similar weighted averages can be calculated for the mean $\mu_{Dj}$ and variance $\sigma^2_{Dj}$ of the $S_p$ sojourn times of cases detected at the $(j+1)^{st}$ screen.

The weighting factors $D_{ij}$ in these expressions are derived by Prorok (1976 a) as a product of probabilities involving the prevalence and incidence rates of preclinical disease in the screened population, the probability that an $i^{th}$ generation individual remains preclinical at least until time $j\Delta$, and the probability that such an individual is detected at time $j\Delta$ but not before. The latter probability involves the quantity $\beta$, the probability that preclinical disease is detected by a screening examination, which is assumed to be constant and identical for each screen.

For purposes of numerical calculation, the underlying disease process is assumed to be stationary over the duration of the screening program so that $q_i(t) = q(t)$ for

all i values. The functional form used for q (t) is the
generalized gamma p.d.f. (see Hager *et al.* (1971)) given by

$$q(t) = \begin{cases} \{\Gamma(r)\ a^{br}\}^{-1}b \quad t^{br-1}\ \exp\ \{-(t/a)^b\}, & 0 \leq t < \infty;\ a,\ b,\ r > 0 \\ 0, & \text{elsewhere.} \end{cases}$$

A relatively short duration disease natural history with a
mean $S_p$ sojourn time of two years is used in the examples.
Results for longer duration disease are qualitatively
similar. Three combinations of the a, b, r parameters
(0.005, 0.3, 5.0; 2.0, 1.0, 1.0; 1.46, 4.5, 4.5) are used to
obtain divergent variances for the q (t) distribution, but
all with a mean of two years. A disease prevalence of 4 per
1000 and incidence of 2 per 1000 per year are assumed.
Screening of each of the three natural histories is
simulated using two values of β, 0.6 and 0.95, and two
screening frequencies, six yearly screens over a five year
period and two screens five years apart.

The three underlying $S_p$ sojourn time p.d.f.s q (t) are
shown in Figure 3. The a, b, r parameter values are given
along with the mean, variance and coefficient of variation
of each distribution. The simulation results for $g_{Dj}(t)$,
$\mu_{Dj}$ and $\sigma^2_{Dj}$ are presented in Figures 4-11 and Tables 1-3.
The Figures display the local density functions for periodic
screening at two different intervals for an underlying
exponential natural history with q (t) parameter values
a = 2.0, b = 1.0, r = 1.0, and for the q (t) density
function having parameters a = 0.005, b = 0.3, r = 5.0. The

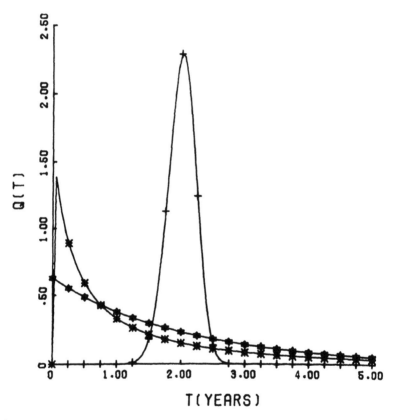

SP SOJOURN TIME P.D.F.

| | A | B | R | MEAN | VAR | CV |
|---|---|---|---|---|---|---|
| ✳ | .005 | .300 | 5.000 | 2.00 | 13.264 | 1.821 |
| + | 1.460 | 4.500 | 4.500 | 2.00 | .047 | .108 |
| ✿ | 2.000 | 1.000 | 1.000 | 2.00 | 4.000 | 1.000 |

Figure 3    Graphs and parameters of three $S_p$ sojourn time
density functions used for numerical simulations.

LOCAL SP SOJOURN TIME P.D.F.S. GSUBDJ(T)

| SP SOJOURN TIME P.D.F. | | | | NO. OF SCREEN | TIMES OF SCREEN | CURVE SYMBOL |
|---|---|---|---|---|---|---|
| A | B | R | MEAN | | | |
| 2.00 | 1.00 | 1.00 | 2.00 | 1.00 | 0 | ☉ |
| | | | | 2.00 | 1.00 | △ |
| BETA = | .60 | | | 3.00 | 2.00 | + |
| | | | | 4.00 | 3.00 | ✳ |
| | | | | 5.00 | 4.00 | ✹ |
| | | | | 6.00 | 5.00 | ◇ |

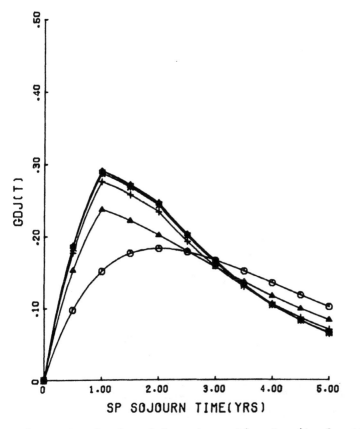

Figure 4    Graphs of $S_p$ sojourn time density functions of cases detected at each screen in a yearly underlying program with $\beta = 0.6$ and an underlying natural history density having parameters a = 2.0, b = 1.0, r = 1.0.

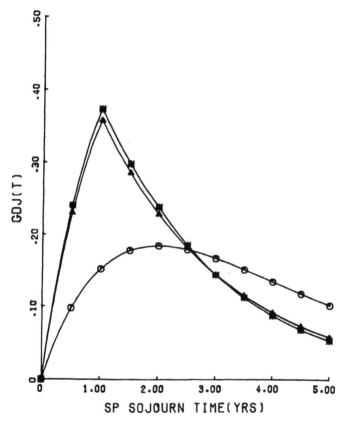

LOCAL SP SOJOURN TIME P.D.F.S. GSUBDJ(T)

SP SOJOURN TIME P.D.F.
| A | B | R | MEAN |
|---|---|---|------|
| 2.00 | 1.00 | 1.00 | 2.00 |

BETA =    .95

| NO. OF SCREEN | TIMES OF SCREEN | CURVE SYMBOL |
|---------------|-----------------|--------------|
| 1.00 | 0 | ☉ |
| 2.00 | 1.00 | △ |
| 3.00 | 2.00 | + |
| 4.00 | 3.00 | ✳ |
| 5.00 | 4.00 | ✿ |
| 6.00 | 5.00 | ◇ |

Figure 5    Graphs of $S_p$ sojourn time density functions of
cases detected at each screen in a yearly
screening program with ß = 0.95 and an underlying
natural history density having parameters a = 2.0,
b = 1.0, r = 1.0.

LOCAL SP SOJOURN TIME P.D.F.S. GSUBDJ(T)

| SP SOJOURN TIME P.D.F. | | | | NO. OF SCREEN | TIMES OF SCREEN | CURVE SYMBOL |
|---|---|---|---|---|---|---|
| A | B | R | MEAN | | | |
| 2.00 | 1.00 | 1.00 | 2.00 | 1.00 | 0 | ⊙ |
| | | | | 2.00 | 5.00 | ▲ |

BETA =    .60

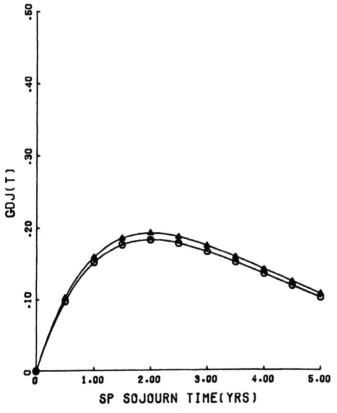

Figure 6    Graphs of $S_p$ sojourn time density functions of
            cases detected at each screen in an every five
            year screening program with β = 0.6 and an
            underlying natural history density having para-
            meters a = 2.0, b = 1.0, r = 1.0.

LOCAL SP SOJOURN TIME P.D.F.S. GSUBDJ(T)

| SP SOJOURN TIME P.D.F. | | | | NO. OF SCREEN | TIMES OF SCREEN | CURVE SYMBOL |
|---|---|---|---|---|---|---|
| A | B | R | MEAN | | | |
| 2.00 | 1.00 | 1.00 | 2.00 | 1.00 | 0 | ⊙ |
|  |  |  |  | 2.00 | 5.00 | ▲ |

BETA =    .95

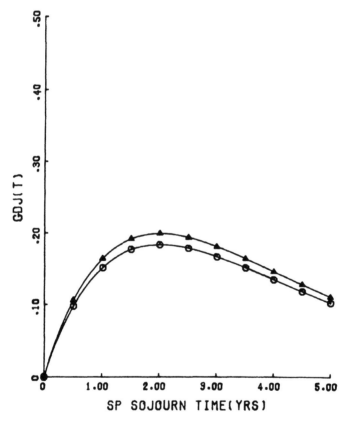

Figure 7    Graphs of $S_p$ sojourn time density functions of
cases detected at each screen in an every five
year screening program with β = 0.95 and an
underlying natural history density having parameters a = 2.0, b = 1.0, r = 1.0.

LOCAL SP SOJOURN TIME P.D.F.S. GSUBDJ(T)

| SP SOJOURN TIME P.D.F. | | | | NO. OF | TIMES OF | CURVE |
|---|---|---|---|---|---|---|
| A | B | R | MEAN | SCREEN | SCREEN | SYMBOL |
| .005 | .300 | 5.000 | 2.00 | 1.00 | 0 | 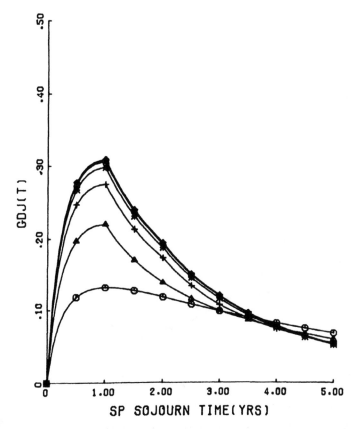 |
| | | | | 2.00 | 1.00 | |
| BETA | = | | .60 | 3.00 | 2.00 | |
| | | | | 4.00 | 3.00 | |
| | | | | 5.00 | 4.00 | |
| | | | | 6.00 | 5.00 | |

Figure 8   Graphs of $S_p$ sojourn time density functions of
cases detected at each screen in a yearly
screening program with $\beta = 0.6$ and an underlying
natural history density having parameters a = 0.005,
b = 0.3, r = 5.0.

LOCAL SP SOJOURN TIME P.D.F.S. GSUBDJ(T)

| SP SOJOURN TIME P.D.F. | | | | NO. OF SCREEN | TIMES OF SCREEN | CURVE SYMBOL |
|---|---|---|---|---|---|---|
| A | B | R | MEAN | 1.00 | 0 | ☉ |
| .005 | .300 | 5.000 | 2.00 | 2.00 | 1.00 | △ |
| | | | | 3.00 | 2.00 | + |
| BETA = | .95 | | | 4.00 | 3.00 | ✻ |
| | | | | 5.00 | 4.00 | ✿ |
| | | | | 6.00 | 5.00 | ◇ |

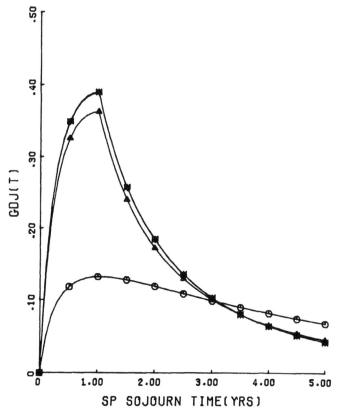

Figure 9    Graphs of $S_p$ sojourn time density functions of cases detected at each screen in a yearly screening program with $\beta = 0.95$ and an underlying natural history density having parameters a = 0.005, b = 0.3, r = 5.0.

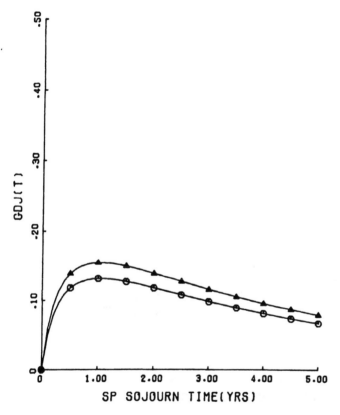

Figure 10   Graphs of $S_p$ sojourn time density functions of
cases detected at each screen in an every five
year screening program with β = 0.6 and an underlying natural history density having parameters
a = 0.005, b = 0.3, r = 5.0.

LOCAL SP SOJOURN TIME P.D.F.S. GSUBDJ(T)

| SP SOJOURN TIME P.D.F. | | | | NO. OF SCREEN | TIMES OF SCREEN | CURVE SYMBOL |
|---|---|---|---|---|---|---|
| A | B | R | MEAN | | | |
| .005 | .300 | 5.000 | 2.00 | 1.00 | 0 | ☉ |
| | | | | 2.00 | 5.00 | △ |

BETA =   .95

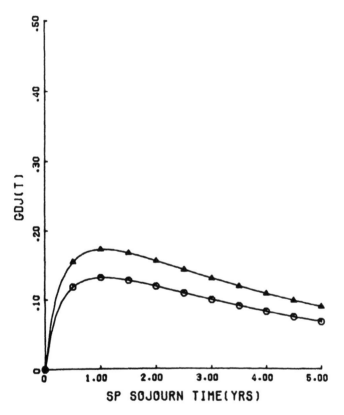

Figure 11    Graphs of $S_p$ sojourn time density functions of cases detected at each screen in an every five year screening program with β = 0.95 and an underlying natural history density having parameters a = 0. 005, b = 0.3, r = 5.0.

Table 1    Local $S_p$ Sojourn Time Means and Variances of
           Individuals Detected by Screening for Natural
           History Parameters a = 2.0, b = 1.0, r = 1.0

| Δ = 1 | β = .6 | | β = .95 | |
|---|---|---|---|---|
| j | $\mu_{Dj}$ (yr) | $\sigma^2_{Dj}$ (yr$^2$) | $\mu_{Dj}$ | $\sigma^2_{Dj}$ |
| 0 | 4.00 | 8.00 | 4.00 | 8.00 |
| 1 | 3.43 | 5.58 | 2.64 | 4.36 |
| 2 | 3.02 | 4.51 | 2.50 | 4.09 |
| 3 | 2.85 | 4.19 | 2.49 | 4.08 |
| 4 | 2.80 | 4.12 | 2.49 | 4.08 |
| 5 | 2.79 | 4.09 | 2.49 | 4.08 |
| Δ = 5 | | | | |
| j | | | | |
| 0 | 4.00 | 8.00 | 4.00 | 8.00 |
| 1 | 3.74 | 5.65 | 3.58 | 5.58 |

legend for each Figure indicates the parameters of the
underlying q (t) density, the value of β, the time of each
screen, and the symbol identifying each curve.  Plots of
$g_{Dj}$ (t) for q (t) having parameter values a = 1.46, b = 4.5,
r = 4.5 are not presented since all the graphs are virtually
identical and very similar to q (t) itself.  Each Table
corresponds to one of the three q (t) density functions and

Table 2    Local $S_p$ Sojourn Time Means and Variances of
           Individuals Detected by Screening for Natural
           History Parameters a = 0.005, b = 0.3, r = 5.0

| Δ = 1 | β = .6 | | β = .95 | |
|:---:|:---:|:---:|:---:|:---:|
| j | $\mu_{Dj}$ (yr) | $\sigma^2_{Dj}$ (yr$^2$) | $\mu_{Dj}$ | $\sigma^2_{Dj}$ |
| 0 | 8.63 | 130.00 | 8.63 | 130.00 |
| 1 | 6.76 | 82.46 | 3.71 | 31.13 |
| 2 | 5.00 | 47.40 | 3.01 | 18.60 |
| 3 | 4.09 | 30.65 | 2.97 | 18.01 |
| 4 | 3.70 | 24.03 | 2.97 | 17.98 |
| 5 | 3.55 | 21.57 | 2.97 | 17.98 |
| Δ = 5 | | | | |
| j | | | | |
| 0 | 8.63 | 130.00 | 8.63 | 130.00 |
| 1 | 6.65 | 57.88 | 5.12 | 34.68 |

contains the local means and variances for that density at
each screen for each of the two screening frequencies and
two values of β.

Some general conclusions about length bias within the
context of this model can be drawn from these calculations.
The magnitude of the length bias effect is seen to vary with
the underlying density q (t), the screening frequency, the
particular screen, and the value of β. The largest effect

Table 3     Local $S_p$ Sojourn Time Means and Variances of
            Individuals Detected by Screening for Natural
            History Parameters a = 1.46, b = 4.5, r = 4.5

| $\Delta = 1$ | $\beta = .6$ | | $\beta = .95$ | |
|:---:|:---:|:---:|:---:|:---:|
| j | $\mu_{Dj}$ (yr) | $\sigma^2_{Dj}$ (yr$^2$) | $\mu_{Dj}$ | $\sigma^2_{Dj}$ |
| 0 | 2.02 | 0.046 | 2.02 | 0.046 |
| 1 | 2.01 | 0.046 | 2.00 | 0.047 |
| 2 | 2.01 | 0.045 | 2.00 | 0.047 |
| 3 | 2.01 | 0.045 | 2.00 | 0.047 |
| 4 | 2.01 | 0.045 | 2.00 | 0.047 |
| 5 | 2.01 | 0.045 | 2.00 | 0.047 |
| $\Delta = 5$ | | | | |
| j | | | | |
| 0 | 2.02 | 0.046 | 2.02 | 0.046 |
| 1 | 2.02 | 0.046 | 2.02 | 0.046 |

occurs at the first screen, the prevalence screen, when $\Delta$ is
small relative to the variation in the natural history
p.d.f. For example, in Table 1 the mean $S_p$ sojourn time of
cases detected at the first screen is 4.00 years, while in
Table 2 it is 8.63 years. Both are considerably larger than
the 2 year mean sojourn time in the underlying natural
history. The variation in the sojourn time distribution of
detected cases is larger than that in q (t) as well.

Although the effect lessens with successive screens, it
still exists at the second and later screens as seen in
Tables 1 and 2 and Figures 4-11. This can be explained by
considering that at the first screen a large proportion of
the longer preclinical duration cases, but relatively few
short duration cases, are available for detection. At
subsequent screens, however, particularly if $\Delta$ is small,
more short duration cases will be detected along with a
preponderance of longer duration cases, so that the overall
$S_p$ sojourn time distribution of detected cases will more
closely resemble the underlying natural history.

The effect of widening the interval between screens is
to make the length bias effect at the second and later
screens for the wider interval more like the effect at the
prevalence screen, provided the wider screening interval is
reasonably large relative to the mean of q (t). This can be
seen in Tables 1 and 2 and by comparing Figures 4 and 6,5
and 7,8 and 10, or 9 and 11. An explanation is that the
larger the value of $\Delta$, the more opportunity the disease has
to recover to its prescreening status before the next screen
interrupts the natural history again. The consequence of
changes in the parameter $\beta$ can be seen in Tables 1 and 2 and
by comparing particularly Figures 4 and 5, or 8 and 9. An
increase in the value of $\beta$, especially to a value close to
its maximum value of one, tends to reduce the length bias
effect more rapidly after the prevalence screen and cause
the effect to be more uniform at screens after the first.

This occurs because if most of the prevalence pool of cases are detected at the first screen as a result of a high β value, then the cases which are detectable at the later screens tend to be a more homogeneous group of primarily incident cases.

The parameter having the largest impact on the magnitude of the length bias effect, however, is the variance of q (t). This is particularly clear if one compares $\mu_{Dj}$ and $\sigma^2_{Dj}$ for the same values of β and Δ across the three Tables. The larger the variance of q (t), the greater the length bias effect. In Table 2, q (t) has large variation and a substantial proportion of very short preclinical duration cases, most of which may become clinical either before screening starts or in an interval between screens, and not be detected by screening. Consequently, the cases that are detected have a considerably larger mean $S_p$ sojourn time than that of the q (t) distribution from which they came. The length bias effect is less but also substantial when q (t) is exponential as seen in Table 1. Also of interest is Table 3. Here the variation in q (t) is very small, indicating that the preclinical duration of the vast majority of cases is very close to the underlying mean of two years. In this circumstance, the length bias effect is negligible, no matter what the value of β or Δ. In fact, in the limiting (but unrealistic) situation where every case of preclinical disease has exactly the same sojourn time, there is no

length bias since the cases cannot be distinguished from one
another on the basis of differences in preclinical state
sojourn time.

## 2. Inadequate Endpoints

Given the necessity of doing a RCT for optimal
evaluation of screening, the choice of outcome measure or
measures must be made. As noted above, the key question of
interest involves determining if screening can favorably
alter the final outcome or event in the disease natural
history; that is, can screening reduce the death rate of the
disease? Consequently, the endpoint of choice is population
mortality. (Functions of this quantity or related measures
which yield the same information may also be appropriate.)
To repeat for emphasis, the mortality rate is defined as the
number of deaths from the disease in question divided by the
number of persons or person-years at risk of dying in the
population under study over some defined time period. The
evaluation should compare the mortality rate in a screened
group with that in a randomized control group at one or more
points in time. This could be done on an age-specific basis
if the study is so designed, or a study could be analyzed in
this way if the data permit.

Of course, this type of evaluation procedure requires
careful study design and long-term followup of large
populations, a usually costly undertaking. Consequently,
alternative short-term outcome measures have been suggested
and the determination of short-term endpoints which are

valid proxys for long-term mortality is an active area of research. There are, however, critical shortcomings associated with endpoints other than mortality which have been suggested. The other endpoints to be considered are more limited than attempting to observe the mortality experience, and the information they yield is correspondingly of more limited value. As stated by Hutchison (1960), such shortcuts are justified only when the relationship between the intermediate and ultimate objectives is established. This relationship has not been established for other endpoints which are mentioned. Additional discussion of endpoint issues appears in Cole and Morrison (1980) and in Prorok *et al.* (1981).

To focus on specific alternative endpoints, consider first the case finding rate. (Similar arguments apply to the incidence rate.) It is certainly of interest to observe this quantity as an early clue to suggest whether or not screening might be doing something. However, this rate yields no information as to the effect of screening on mortality. One would expect this rate to increase, at least initially, relative to an unscreened population because of lead time bias, but this can happen in the presence or absence of a mortality effect. Furthermore, one must be careful about what is being counted as discovered cancer in an early detection program. Many of the borderline lesions found by newer screening modalities may not be disease at all. A particularly striking example of the increase in

incidence as a result of the introduction of a screening
program is presented in the Walton Report (1976) as part of
a discussion of cervical cancer screening in Saskatchewan.

Stage of disease at diagnosis is another criterion that
has been offered for the evaluation of screening. This too
should be examined as an early indicator that screening
might be accomplishing something, but it is unsatisfactory
as a final endpoint for a number of reasons. The definition
of stage can be relatively subjective and its proper use
requires strict guidelines and tight control over pathology,
each of which may be difficult to implement in practice.
Most importantly, the relationship of stage to survival or
mortality in the screening setting is not established. What
extent of shift in stage distribution of cases as a result
of screening corresponds to what size reduction in
mortality? How does the detection of in situ or borderline
cases affect stage distribution and subsequent mortality?
The data to answer these questions are not available at this
time.

The problem is likely to be most pronounced for Stage I
or localized cases, particularly if the study has a cut-off
point after which new cases are not accrued. In this
circumstance, lead time and length bias can result in slow
growing, even nonprogressive, cases being detected in Stage
I in the screened group to a greater extent than in the
control group. Their counterparts in the control group may
not surface by the cut-off-point, if they ever surface, and

as a result the screened group will contain a higher
proportion of Stage I cases even if screening has no effect
on mortality. If there is a mortality effect, the magnitude
could be exaggerated by confining an analysis to stage of
disease.

Additionally, if one wants to use stage and relate this
somehow to survival, it must be recognized that even within
a given stage, survival in a screened group can be different
than in a control group because of length bias.
Alternatively, in the HIP study, among the interval cases
(detected between screens, not as a result of screening),
about 58% had no axillary nodal involvement. In the control
group, 46% of the cases had no nodal involvement, a
substantial difference. Yet the survival rates a 8 years
were very close, 49% and 50%, respectively. Thus an
examination of stage alone could lead to a wrong conclusion
about prognosis as discussed by Beahrs *et al*. (1979).

Furthermore, in using mortality data to evaluate
screening, it should be standard procedure to collect and
compare data on total mortality as well as mortality from
the disease of interest. If stage were to be used as an
endpoint, an analogous procedure would be necessary. That
is, one would have to collect stage information on relevant
major diseases in both the study and control groups in order
to fully analyze the potential selection biases mentioned in
Section II B.

The final sub-optimal endpoint to be considered is
survival, specifically the survival rate. In contrast to

mortality, which is a population measure, survival refers only to cases within a population. The N-year survival rate is defined as the number of cases alive after N years of observation divided by the number of cases diagnosed at the beginning of the time period. Of course there are losses to followup and this measure is ordinarily calculated using life table methods. While this variable does address the final outcome of disease and should be examined for suggestive evidence of screening effectiveness, it cannot be relied upon to accurately reflect mortality for reasons which center on the lead time and length biases and the lack of methodology to completely adjust for these biases.

The objective of a cancer screening program is to detect and treat disease earlier than usual and thereby prevent or postpone death from the cancer. If this in fact does happen, it should be reflected in an increased survival rate as well as a mortality reduction. However, since diagnosis also occurs earlier in a screening program, any observed increase in survival from time of diagnosis is, at least in part, simply a reflection of lead time. A pertinent illustration is given by Zelen (1976). Suppose an individual enters $S_p$ at age 50, the disease metastasizes at age 55 and causes death at age 62. Without screening, symptoms would lead to diagnosis at age 60, so that the survival time is 2 years. If the person had been screened, the disease may have been found at age 58, but since it had already metastasized, death would still occur at age 62.

The observed survival would now be 4 years, but the person
died at the same age so that no therapeutic benefit was
realized from screening. The artificial 2 year survival
increase is simply lead time. Of course, it may also
happen that a similar individual is diagnosed by screening
at age 54, before metastasis, and then lives until age 65,
for example. This person's observed survival is 11 years
with a lead time of 6 years and a real benefit of three
extra years of life. In an actual screening program, both
situations may occur, but it is impossible to distinguish
them because lead time cannot be directly observed for
ethical reasons. Consequently, the use of survival to
measure screening effectiveness is open to question.

One could attempt to compare the survival rate in a
screened group with that in a randomized control group after
adjusting the screened group survival for lead time.
Unfortunately, no completely satisfactory adjustment
procedure is available as discussed by Thomas *et al.* (1977).
The suggested adjustment procedures require knowledge of the
average lead time. Methods for estimating this quantity
have been suggested (see Hutchison and Shapiro (1968),
Shapiro *et al.* (1974), Zelen and Feinleib (1969), and Prorok
(1976 b), but the problem is not completely solved,
particularly for repeated screening (see Thomas *et
al.* (1977)). The problem becomes even more complicated when
one attempts to compare survival between subgroups, such as
age groups or groups of cases detected by different

screening modalities.  The  lead time effect may very well
be different in different subgroups, yet no satisfactory
procedure is available to estimate the mean  lead time
within subgroups or to adjust subgroup survival for  lead
time.

Even if one could adjust for  lead time, the problem of
length bias would still exist in making survival
comparisons, particularly between subgroups.  For example,
the length bias effect may be different between two
subgroups of cases detected by different screening
modalities.  That is, the cases in one subgroup may have a
different distribution of natural histories than the cases
in another subgroup because of a modality dependent sampling
effect.  Thus even if one could adjust for  lead time, any
remaining survival difference could be real, or it could
simply be a consequence of the difference in disease natural
history between the two groups, or a combination of the two
factors.  Unfortunately, as discussed by Thomas *et al*.
(1977), no general methodology exists to either estimate the
magnitude of a length bias effect or adjust survival for
length bias.  An approach to separating the effects of
treatment,  lead time and length bias in certain
circumstances has been proposed by Morrison (1982).


III.   STUDY DESIGN FOR THE RIGOROUS EVALUATION OF SCREENING
       PROGRAMS

Several authors, including Lave and Lave (1978) and
Henderson and Meinert (1975), have reflected on the

evaluation of health care programs in our society.
Approaches have been outlined for determining both cost and
benefit of such programs, with the recognition that despite
the difficulties involved in accurate evaluation, such
activities should be improved and applied more widely rather
than curtailed. Billions of dollars are spent every year on
health care and some, perhaps substantial, part of that sum
will be wasted unless unresolved questions are answered as
accurately as possible using some well thought out,
systematic approach. A recent task force on the economic
impact of preventive medicine reached certain conclusions
relevant to screening as well as other health care
maneuvers. Their report indicated that health care programs
must be evaluated if we are to realize the potential of
medical care and contain cost, and that therapeutic and
preventive health care programs should be evaluated by
equally critical techniques and criteria. In addition,
research must be accomplished to determine the efficacy and
cost of proposed programs, and the large scale clinical
trial is one of the most important techniques for developing
evidence on efficacy.

The need for evaluation exists for many health
procedures such as use of chemotherapeutic agents,
procedures in coronary care units, new surgical techniques
and health maintenance programs, in addition to screening.
The prime tool for such evaluation on the local, national or
international level is the clinical trial as described by
Henderson and Meinert (1975). The design and conduct of

such trials can be difficult, time consuming and costly.  It
is therefore useful to review certain design and
implementation considerations for evaluation trials before
outlining certain basic study designs.  First, the specific
problem to be studied must be identified and the outcome
variable must be determined.  Choice of the sampling unit is
a key design factor.  The individual is the unit of choice
in most circumstances, but the family unit or entire
communities might be more appropriate in certain
circumstances.  The admission and exclusion criteria should
be clearly established.  A critical part of any evaluation
effort is the selection of a control group.  This should be
done using a bona fide randomization procedure, and control
and study groups should be followed with equal intensity in
the same time frame.  The randomization procedure may be
augmented by stratification which must then be taken into
account during data analysis.  Efforts should be made to
ensure adherence to the study protocol and the degree of
adherence should be reported.  Administration of treatment
should be done in double-blind fashion where possible.  In
screening trials, this is probably not feasible, but it is
possible and prudent to measure the outcome variable, namely
life or death in cancer screening, using a blinded
procedure.  Cost factors should be given serious scrutiny
before and during a study.  These are mainly related to
recruitment of study subjects, administration of the test or
treatment and management of patients, and data collection

and processing. In reference to the latter, serious efforts should be made to limit data collection to those variables relating to the main aims of the study.

Screening for the early detection of cancer involves intervention into a well population at the initiation of the screener, not the screenee, with some implied promise of benefit. It involves real and potential risks and may be very costly. There is a requirement to quantify benefit and cost, rather than settle for "yes-no" answers. This type of health maneuver demands rigorous and unemotional evaluation. As indicated by the discussion above and in the previous parts of this chapter, the only proper way to accomplish this evaluation is the randomized controlled trial with mortality as the endpoint (RCTM). These involve the prospective testing and followup of defined populations according to a predetermined protocol for an extended time period, usually at least 10 years. Given the effort involved, a large number of such trials cannot realistically be contemplated. However, the RCTM should be used to address certain basic questions for each major cancer site for which a good screening test exists, including benefit, optimal frequency, and marginal contribution of one screening modality when two or more are available. A brief outline of possible study designs to address these questions follows.

A benefit RCTM is designed to yield definitive scientific evidence as to the benefit of screening compared

to the usual medical care available to the population under
study. The eligible population is randomized into two
groups, the study group and the control group. The
screening test is applied or offered to the study group at
some appropriate frequency over a certain period of time
while the control group receives its usual medical care over
the same time period. The study and control group mortality
rates are then compared after some appropriate followup
period. The HIP breast cancer screening trial is a RCTM
benefit study aimed at determining the effect on breast
cancer mortality of screening with the combination of
mammography and clinical examination (see Shapiro (1977) and
Shapiro et al. (1982)). A benefit RCTM in lung cancer is
being conducted at the Mayo Clinic using the combined
modalities of Xray and sputum cytology (see Taylor and
Fontana (1972) and Taylor et al. (1981)).

A frequency RCTM is aimed at comparing the impact of
different screening frequencies on reducing disease
mortality, thereby contributing information crucial to the
determination of an optimal screening frequency. The
eligible population is randomized into two (or more) study
groups. One group is screened at one frequency, say every
year, while the other group is screened at another
frequency, say every two years. The death rates in the two
groups are then compared after an appropriate followup
period. A RCTM sponsored by the National Cancer Institute
using occult blood in the stool as a screening test for

colon cancer is in part designed to address this frequency question. One randomized group is screened every year while another is screened every two years. There is a third study group, a control group, so that the benefit question can be studied as well (see Gilbertsen *et al.* (1980)).

The objective of a marginal contribution RCTM is to ascertain the independent or additional mortality reduction which can be attributed to a given screening test when more than one test is available. Suppose two screening tests, $S_1$ and $S_2$, are available, with $S_1$ being considered the standard test and $S_2$ a new test. The eligible population is randomized into two groups. One group is screened with test $S_1$ at some appropriate frequency while the other group is administered both $S_1$ and $S_2$ at the same frequency. The death rates in the two groups are then compared after an appropriate followup period to determine the additional benefit to be derived by using $S_2$ in combination with $S_1$. A RCTM of this type to determine the independent contribution of mammography in screening women for breast cancer is part of the current Canadian breast cancer study (see Miller (1982)).

It should be noted that the appropriate screening frequency to use in a benefit study, the frequencies to compare in a frequency study, the number of screens to use in these studies and the length of the followup period all must be decided upon in individual situations. An established general scheme for making these decisions is not available but is the subject of current research.

It may happen in certain settings that the outlined designs are not applicable. Certain modifications which maintain the basic integrity of the evaluation may then be considered. For example, if one is interested in determining screening benefit in an industrial population or a very high risk group, it may not be feasible for the control group to receive only its usual medical care. The proximity of and communication between employees who work together every day but who are randomized to different groups could lead to severe contamination problems and compromise a study. Alternatively, certain investigators may feel that very high risk individuals must be offered more than the usual medical care that a control group would ordinarily have available to it. In these circumstances, a possible alternative RCTM design, proposed by Zelen (1974), involves screening for two diseases at the same time. That is, one randomized group is screened for one disease and the other group is screened for a different disease. Each group serves as a control for the other. Possible disease pairs include breast cancer and cervical cancer, lung cancer and bladder cancer, or cancer and heart disease. The problems and implications associated with this design have not been investigated as fully as in the simpler designs discussed previously and caution is advised in using this modified approach. For example, population characteristics associated with high risk of breast and cervical cancer are quite different, so that a breast cancer-cervical cancer pairing might not be workable in practice.

An alternative RCTM modification involves changing the unit of randomization from the individual to a larger group such as the family, a section of a factory, an entire factory or a community. Use of larger units may reduce the statistical power of the study or necessitate the followup of larger populations, but costs may also be reduced and ethical problems obviated as discussed by Henderson and Meinert (1975). Matching or stratification of units may be more important in this modified design to help maintain statistical power than is the case when the individual is the randomization unit. A situation where this approach may be relevant is screening for cervical cancer. A randomized trial to evaluate cervical cancer screening has never been done so that unequivocal evidence of the benefit of cytologic screening does not exist. Nonetheless, many in the medical community believe the procedure is of value and consequently a benefit RCTM as described above is probably not feasible. Details of a study design using census tract as the unit of randomization rather than the individual to evaluate cervical cancer screening have been reported (see Apostolides and Henderson (1977)). As a further example, an international RCT was organized by the World Health Organization (1974) to estimate the extent to which the main risk factors for coronary heart disease (CHD) can be modified in industrial workers and the effect of such changes on CHD incidence and mortality. The allocation units in this trial were factories or other large occupational units.

IV.  SUBSIDIARY EVALUATION PROCEDURES

In those instances where it is effectively impossible to
perform a RCTM, certain subsidiary approaches to evaluation
of cancer screening must be used.  Further, these sub-
optimal approaches may serve as adjuncts for addressing
scientific questions even when a RCTM has been implemented.
Two subsidiary approaches will be briefly discussed,
observational studies and mathematical modelling.

   An observational study must be well designed and
carefully conducted if it is to be at all useful for
evaluation purposes.  The criteria for a good evaluation
study listed in Section II should be met by an observational
study just as for an experimental study, save for
randomization.  The best observational study would likely
focus on a comparison of cancer incidence and mortality in a
defined population before and after the introduction of a
screening program.  Time trends in incidence and mortality
would be examined and inter-area comparisons of intensively
screened areas with non-screened areas could be made.  This
approach would require rapid introduction of the screening
program and virtually full coverage of the population at
risk.  Reliable incidence and mortality data from at least
10 years prior to the start of screening and which is
predictable for the future must be available.  Ideally, this
would be total incidence and mortality data, not simply for
the cancer of interest.  A clear requirement is the
capability for accurate, long-term followup of the entire

population at risk. Such an observational study may require substantially larger populations than the RCTM and may eventually prove more costly than the more rigorous design (see Miller (1978 b)).

The observational approach just described has been advocated as a means to determine the benefit of cytologic screening for the early detection of cervical cancer. Unfortunately, virtually none of the existing studies satisfy all the important design criteria. One possible exception is the cervical cancer screening program in Iceland described by Johannesson *et al.* (1978). Another possible application of this type of study is as a confirmation trial (see Miller (1978 c)). That is, once a given question, such as screening benefit, has been answered affirmatively by a RCTM, it may be useful to confirm or replicate the result by one or more observational studies in the situation where further RCTMs are not feasible. It should be mentioned that these applications of observational studies in cancer screening have been attempted in only a few instances and the methodology of observational studies is in need of further development.

One can easily imagine a multitude of questions to be answered about screening programs, particularly with regard to various age ranges of the population and screening frequencies. It is likely that only the most important questions will be answered by evaluation studies, including perhaps one benefit study, one frequency study and one or two marginal contribution studies for each major cancer. A less expensive and time consuming alternative may be necessary to address the many remaining questions. One possibility is to use mathematical models, along with available data, to simulate various screening program options and predict cost and benefit. Considerable progress has been made in recent years in the development of such models including those referenced in previous sections and others described in Dubin (1981), Eddy (1980), Kirch and Klein (1978), and Knox (1973). A discussion of modelling concepts appears in Chapter 8 of this book by Shwartz and Plough.

Models have essentially three uses in the evaluation of cancer screening. Simulation results may serve to substitute for study results when evaluation trials cannot be done. Further, simulation results can aid in the design of evaluation trials by narrowing the range of reasonable options to be entertained for such trials. For example, the range of possible screening frequencies might be narrowed to two or three yielding potential benefits of public health importance yet different enough to be compared using a feasible study design. Finally, modelling can yield new methodology and provide a methodologic framework for the analysis of data from screening studies (see Louis (1978)). However, before model output can be relied upon for these purposes, the model must be validated; that is, it must somehow be shown to be consistent with good quality experimental data on the benefits and costs of particular screening protocols for diseases of interest. The good quality data necessary for validation can be obtained from randomized controlled trials.

## ACKNOWLEDGMENTS

The author is indebted to Kathleen Graff and Joyce Campbell for assistance in preparation of the Figures, and to Carol Ball and Betty Hennigan for typing the manuscript.

## REFERENCES

Albert, A., Gertman, P., and Louis, T. (1978 a). Screening for the early detection of cancer - I. The temporal natural history of a progressive disease state. *Mathematical Biosciences*, 40: 1-59.

Albert, A., Gertman, P., Louis, T., and Liu, S. (1978 b). Screening for the early detection of cancer - II. The impact of screening on the natural history of the disease. *Mathematical Biosciences*, 40: 61-109.

Apostolides, A. and Henderson, M. (1977). Evaluation of cancer screening programs, parallels with clinical trials. *Cancer*, 39: 1779-1785.

Bailar, J.C. (1976). Mammography: A contrary view. *Annals of Internal Medicine*, 84: 77-84.

Beahrs, O.H. *et al.* (1979). Report of the working group to review the National Cancer Institute - American Cancer Society breast cancer detection demonstration projects. *Journal of the National Cancer Institute*, 62: 639-709.

Blumenson, L.E. (1976). When is screening effective in reducing the death rate? *Mathematical Biosciences*, 30: 273-303.

Blumenson, L.E. (1977). Detection of disease with periodic screening: Transient analysis and application to mammography examination. *Mathematical Biosciences*, 33: 73-106.

Buck, A.A. and Gart, J.J. (1966). Comparison of a screening
    test and a reference test in epidemiologic studies:
    I.  Indices of agreement and their relation to
    prevalence. *American Journal of Epidemiology*, *83*:
    586-592.

Byar, D.P., Simon, R.M., Friedewald, W.T., Schlesselman,
    J.J., DeMets, D.L., Ellenberg, J.H., Gail, M.H., and
    Ware, J.H. (1976).  Randomized clinical trials:
    Perspectives on some recent ideas. *New England
    Journal of Medicine*, *295*: 74-80.

Cole, P. and Morrison, A.S. (1980).  Basic issues in
    population screening for cancer. *Journal of the
    National Cancer Institute*, *64*: 1263-1272.

Dubin, N. (1981).  Predicting the benefit of screening for
    disease. *Journal of Applied Probability*, *18*:
    348-360.

Eddy, D.M. (1980).  *Screening for Cancer - Theory, Analysis
    and Design*.  Prentice-Hall, Englewood Cliffs, New
    Jersey.

Galen, R.S. and Gambino, S.R. (1975).  *Beyond Normality: The
    Predictive Value and Efficiency of Medical
    Diagnosis*.  Wiley, New York.

Gart, J.J. and Buck, A.A. (1966).  Comparison of a screening
    test and a reference test in epidemiologic studies:
    II.  A probabilistic model for the comparison of
    diagnostic tests. *American Journal of Epidemiology*,
    *83*: 593-602.

Gilbertsen, V.A., Church, T.R., Grewe, F.J., Mandel, J.S.,
    McHugh, R.B., Schuman, L.M., and Williams,
    S.E. (1980).  The design of a study to assess
    occult-blood screening for colon cancer. *Journal of
    Chronic Diseases*, *33*: 107-114.

Hager, H.W., Bain, L.J., and Antle, C.E. (1971). Reliability
    estimation for the generalized gamma distribution
    and robustness of the Weibull model. *Technometrics*,
    *13*: 547-557.

Henderson, M. (1976).  Validity of screening. *Cancer,
    Supplement*, *37*: 573-581.

Henderson, M.M. and Meinert, C.L. (1975).  A plea for a
    discipline of health and medical evaluation.
    *International Journal of Epidemiology*, *4*: 11-23.

Herbert V. (1977).  Acquiring new information while
    retaining old ethics. *Science*, *198*: 690-693.

Hutchison, G.B. (1960). Evaluation of preventive services. *Journal of Chronic Disease*, 11: 497-508.

Hutchison, G.B. and Shapiro, S. (1968). Lead time gained by diagnostic screening for breast cancer. *Journal of the National Cancer Institute*, 41: 665-681.

Johannesson, G., Geirsson, G., and Day, N. (1978). The effect of mass screening in Iceland, 1965-1974, on the incidence and mortality of cervical carcinoma. *International Journal of Cancer*, 21: 418-425.

Kirch, R.L.A. and Klein, M. (1978). Prospective evaluation of periodic breast examination programs: Interval cases. *Cancer*, 41: 728-736.

Knox, E.G. (1973). A simulation system for screening procedures. In *The Future and Present Indicatives, Problems and Progress in Medical Care, Ninth Series* (G. McLachlan, ed.). Nuffield Provincial Hospitals Trust, Oxford, London, pp. 17-55.

Lave, J.R. and Lave, L.B. (1978). Cost-benefit concepts in health: Examination of some prevention efforts. *Preventive Medicine*, 7: 414-423.

Lilienfeld, A.M. (1974). Some limitations and problems of screening for cancer. *Cancer, Supplement, 33*: 1720-1724.

Louis, T.A., Albert, A., and Heghinian, S. (1978). Screening for the early detection of cancer - III. Estimation of disease natural history. *Mathematical Biosciences*, 40: 111-144.

Miller, A.B. (ed.) (1978 a). Report of discussion on general principles of screening. *Screening in Cancer: A Report of a UICC Workshop Toronto Canada, April 24-27, 1978*. International Union Against Cancer, Geneva, pp. 64-70.

Miller, A.B. (ed.) (1978 b). Summary and general recommendations. *Screening in Cancer. A Report of a UICC Workshop, Toronto Canada, April 24-27, 1978*. International Union Against Cancer, Geneva, pp. 334-338.

Miller, A.B. (1978 c). Risk benefit in mass screening programs for breast cancer. *Seminars in Oncology*, 5: 351-359.

Miller, A.B. (1982). Evaluation of screening for cancer of the cervix and breast. Implications for cancer control. In *Issues in Cancer Screening and*

*Communications* (C. Mettlin and G.P. Murphy, eds.). Alan R. Lus, New York, pp. 41-54.

Morrison, A.S. (1982). The effects of early treatment, lead time and length bias on the mortality experienced by cases detected by screening. *International Journal of Epidemiology*, 11: 261-267.

Peto, R., Pike, M.C., Armitage, P., Breslow, N.E., Cox, D.R., Howard, S.V., Mantel, N., McPherson, K., Peto, J., and Smith, P.G. (1976). Design and analysis of randomized clinical trials requiring prolonged observation of each patient. I. Introduction and design. *British Journal of Cancer*, 34: 585-612.

Peto, R., Pike, M.C., Armitage, P., Breslow, N.E., Cox, D.R., Howard, S.V., Mantel, N., McPherson, K., Peto, J., and Smith, P.G. (1976). Design and analysis of randomized clinical trials requiring prolonged observation of each patient. II. Analysis and examples. *British Journal of Cancer*, 35: 1-39.

Prorok, P.C. (1976 a). The theory of periodic screening I: Lead time and proportion detected. *Advances in Applied Probability*, 8: 127-143.

Prorok, P.C. (1976 b). The theory of periodic screening II: Doubly bounded recurrence times and mean lead time and detection probability estimation. *Advances in Applied Probability*, 8: 460-476.

Prorok, P.C., Hankey, B.F., and Bundy, B.N. (1981). Concepts and problems in the evaluation of screening programs. *Journal of Chronic Diseases*, 34: 159-171.

Sackett, D.L. (1975). Periodic examination of patients at risk. In *Cancer Epidemiology and Prevention, Current Concepts* (D. Schottenfeld, ed.). Charles C. Thomas, Springfield, Illinois, pp. 437-454.

Sackett, D.L. and Holland, W.W. (1975). Controversy in the detection of disease. *The Lancet*, August 23, 1975, 357-359.

Schneiderman, M.A. (1966). Looking backward: It is worth the crick in the neck? Or: Pitfalls in using retrospective data. *American Journal of Roentgenology, Radium Therapy and Nuclear Medicine*, 96: 230-235.

Screening for Disease. A series from *The Lancet*, October 5 - December 21, 1974.

Shapiro, S. (1977). Evidence on screening for breast cancer
from a randomized trial. *Cancer, Supplement, 39*:
2772-2782.

Shapiro, S., Goldberg, J.D., and Hutchison, G.B. (1974).
Lead time in breast cancer detection and
implications for periodicity of screening. *American
Journal of Epidemiology, 100*: 357-366.

Shapiro, S., Venet, W., Strax, P., Venet, L., and Roeser,
R. (1982). Ten-to fourteen-year effect of screening
on breast cancer mortality. *Journal of the National
Cancer Institute, 69*: 349-355.

Shwartz, M. (1978 a). A mathematical model used to analyze
breast cancer screening strategies. *Operations
Research, 26*: 937-955.

Shwartz, M. (1978 b). An analysis of the benefits of serial
screening for breast cancer based upon a
mathematical model of the disease. *Cancer, 41*:
1550-1564.

Taylor, W.F. and Fontana, R.S. (1972). Biometric design of
the Mayo Lung Project for early detection and
localization of bronchogenic carcinoma. *Cancer, 30*:
1344-1347.

Taylor, W.F., Fontana, R.S., Uhlenhopp, M.A., and Davis,
C.S. (1981). Some results of screening for early
lung cancer. *Cancer, 47*: 1114-1120.

Thomas, L.B. *et al.* (1977). Report of NCI ad hoc pathology
working group to review the gross and microscopic
findings of breast cancer cases in the HIP Study
(Health Insurance Plan of Greater New York).
*Journal of the National Cancer Institute, 59*:
495-541.

Thorner, R.M. and Remein, Q.R. (1961). *Principles and
Procedures in the Evaluation of Disease.* Public
Health Monograph No. 67, U.S. Public Health Service,
Washington, D.C.

Walton, R.J. *et al.* (1976). Cervical cancer screening
programs. *Canadian Medical Association Journal,
114*: 1003-1033.

Wilson, J.M.G. and Jungner, G. (1968). *Principles and
Practice of Screening for Disease.* Public Health
Paper No. 34, World Health Organization, Geneva.

World Health Organization European Collaborative
    Group. (1974).   An international controlled trial in
    the multifactorial prevention of coronary heart
    disease.   *International Journal of Epidemiology*, 3:
    219-224.

Zelen, M. (1974).   Personal communication.

Zelen, M. (1976).   Theory of early detection of breast
    cancer in the general population.   In *Breast Cancer:
    Trends in Research and Treatment* (J.C. Heuson,
    W.H. Mattheiem, and M. Rozencweig, Eds.).   Raven
    Press, New York, pp. 287-300.

Zelen, M. and Feinleib, M. (1969).   On the theory of
    screening for chronic diseases.   *Biometrika*, 56:
    601-614.

8
# MODELS TO AID IN PLANNING
# CANCER SCREENING PROGRAMS

Michael Shwartz
Boston University
Boston, Massachusetts

Alonzo L. Plough
Tufts University
Medford, Massachusetts

## I. INTRODUCTION

The development of a comprehensive and valid strategy for
cancer control is the underlying goal of all forms of cancer
research.  In the light of limited advances toward a "cure"
in basic biomedical research, research has recently focused
on the issue of prevention.  However, because the etiologies
of cancer have been, and to a large extent remain,
incompletely understood, the early detection of cancer has
become an important part of this focus.  Thus, screening

        Preparation of this manuscript was supported in part
by Grant R18 CA 17807-04 awarded by the National Cancer
Institute, DHEW to Boston University.

programs are playing a larger role in the cancer control
effort.

As discussed in Chapter 7, randomized controlled trials
(rct's) provide the most satisfactory empirical basis for
determining the gains than can be realized from the earlier
detection of different types of cancer.  However, though
rct's may provide the justification for screening, the
design of an actual screening program in any given situation
raises more questions than can be answered by rct's since
decisions about program design involve a joint consideration
of population-specific characteristics and resource
constraints.  Therefore, the questions addressed in any
particular rct must be considered anew when population
characteristics, disease characteristics, screening
technologies or resource constraints are substantially
different from the "model" rct.

For example, from an rct we can determine that if some
particular population (a study group) is offered screening
with a certain modality at a particular frequency, benefits
result (or do not result) when the study group is compared
with some group that is not offered screening (the control
group).  However, in the design of any particular screening
program we must consider that our population may be quite
different than that in the rct.  The target population may
have a different age distribution, be exposed to different
risk factors, and have specific behavioral characteristics
such that its response to disease symptoms and potential

participation in a screening program diverge significantly
from that observed in the rct.

Because of the expense of randomized controlled trials
and the long time necessary to complete such studies, it is
clearly impossible to perform trials to evaluate all program
options in all types of settings. Also, some screening
techniques have been adopted before rct's have been
performed (e.g., the Pap smear). In these cases, it is
necessary to design programs with even less complete
information. In the occupational environment, rct's are
particularly difficult because of small sample sizes, unique
population characteristics, and the inability to easily
monitor individual exposures to a wide variety of substances
in the work place. Because of the importance of
occupational exposures that result in excess risk for
cancer, methods must be developed to quantitatively evaluate
various control strategies in industrial populations when an
rct may not be feasible.

For these types of reasons, it is most important to
have some basis for using available data to aid a decision-
maker in considering the range of questions that arise in
the design of an actual screening program - be it in a
general population or an industrial population. It is our
contention that mathematical models can be extremely useful
in aiding in the design of screening. Given even quite
limited information, mathematical models can be used to
evaluate a wide range of possible approaches to screening

programs that have in fact not yet been implemented. Thus, models provide an explicit basis for extrapolating from limited experience.

In the first part of this chapter, we will discuss mathematical models that have been developed to analyze screening strategies in the general population. In order to provide some insight into model structure, parameter estimation and model verification, we will discuss a simple model that could be used to evaluate screening strategies. We will than consider general formulations that have been proposed to analyze cancer screening strategies and finally, discuss the actual analysis of screening strategies for breast, cervical and colo-rectal cancer.

The second part of the chapter will focus on screening in industrial populations. Whereas over the last 15 years significant work has been done in developing models to aid in analyzing general population screening strategies, relatively little emphasis as been given to screening in industrial populations. Also, because of the nature of the cancer problem in industrial populations, there appears significantly more potential for cancer control activities that focus on identifying and limiting exposure to carcinogens. This potential suggests a shift in emphasis from screening people to screening substances. We will discuss issues in screening substances for carcinogenicity and then illustrate an approach for analyzing screening strategies for substances.

II.  MODELS TO ANALYZE SCREENING STRATEGIES IN THE GENERAL
      POPULATION

In general, there are two types of models that have been
used to analyze cancer screening strategies.  These two
types are perhaps most clearly distinguished by Bross *et al.*
(1968), who call one type "surface models" and the other
type "deep models."  The difference in the two is the types
of events that are emphasized in the analysis.  Surface
models,which comprise the usual statistical approach to
analysis, consider only those events that can be directly
observed, e.g., disease incidence, prevalence and mortality.
Deep models, on the other hand, emphasize those events which
cannot be directly observed, i.e., they incorporate
hypotheses about the underlying disease process that
generates the observable events.

     The intent of a deep model is not just to account for
any specific set of surface events, but to use surface
events as a basis for understanding underlying dynamics.
This understanding, as incorporated in a series of
hypotheses, permits generalization from the particular
experiment that generated the surface events.  As a result,
whereas surface models provide a basis for interpreting the
observable effects of screening, deep models provide an
explicit basis for determining the outcomes of screening
scenarios that have not been studied in clinical trials.
For this reason, deep models can be particularly powerful
aids to policy makers.

In order to develop a deep model, some degree of understanding of the underlying disease process is necessary. For several cancers of high prevalence in the general population, sufficient information does exists to develop deep models. Because of the usefulness of deep models as a means of extrapolating from observable results to untested strategies, we will focus on these models in this chapter.

In the next section, we present, for a simplified deep model, the mathematical equations that could be used to evaluate alternative screening strategies and briefly discuss model parameterization and validation. For those who find mathematical equations enlightening, this section should provide a better understanding of the probabilistic foundation of disease models. In the second section, we discuss models that have been developed to analyze cancer screening strategies. This section can be read independently of the first section.

## A. A Simple "Deep" Model to Evaluate Screening Strategies

### 1. *Overview*

Assume a person can be in one of three states: healthy, preclinical or clinical. A person enters the preclinical state when cancer develops, where development may be defined either as the time of carcinogenesis, or more usually as the time when the disease can first be detected by a screening modality. For example, in the case of breast cancer, a

woman might be assumed to enter the preclinical state when
her tumor becomes .5 cm in diameter, the smallest tumor size
that can regularly be detected by mammography; in the case
of cervical cancer, a woman might enter the preclinical
state when carcinoma in situ develops.

Once a person enters the preclinical state, it is
assumed that the disease will progress in such a way that as
the length of time the person has had the disease before
treatment increases, prognosis at the time of treatment
declines. This relationship between the duration of
preclinical disease and prognosis is a fundamental
hypothesis that underlies any screening effort. For, to the
extent there is little relationship between prognosis and
the duration of preclinical disease, there is little value
in earlier detection and hence, little value in screening.

As a person ages, there is some probability at each age
of entering the preclinical state, i.e., of making a
transition from healthy to preclinical disease. We will
refer to this probability as the age-specific rate of the
disease development. At some time in the course of disease
progression, the disease will come to clinical attention in
the absence of participation in any screening regimen. This
is the event of transition from preclinical to clinical
disease. We will refer to this event as clinical surfacing.
To reflect both biological variability and uncertainty about
the rate of disease progression in any individual, clinical
surfacing is considered a stochastic event.

If a person is screened while in the preclinical state, the disease may be detected, depending on the reliability of the screening technique. If detected, the preclinical state will be shorter than if the disease has clinically surfaced. Benefits from screening will result to the extent the disease is detected earlier and to the extent prognosis is more favorable the shorter the duration of preclinical disease.

## 2. *Mathematical Formulation*

To illustrate this model, we present below the types of equations that could be used to evaluate a given screening strategy. For this illustration, it is useful to define a hazard rate. Let T be a random variable representing the time of occurrence of some event and $f(t)$ be the probability density function for the random variable T. Then

$$F'(t) = \int_t^\infty f(x)dx$$

equals the probability the event has not occurred by time t. A hazard rate $r(t)\Delta t$ is defined as $f(t)\Delta t/F'(t)$, which equals the probability the event occurs during some small interval $\Delta t$ following time t, given that the event has not occurred by time t. Given $r(t)$, it can be shown that

$$F'(t) = \exp(-\int_0^t r(u)du); \quad f(t) = r(t) \exp(-\int_0^t r(u)du).$$

We will denote a hazard rate by a small letter, e.g. $r(t)$.

The corresponding capital letter, e.g. R(t), will denote

$$\int_0^t r(u)du.$$

Thus, a probability density function can be written as $r(t) \exp(-R(t))$.

To evaluate a screening strategy, we need the following:

$i_r(t)$ = hazard rate for disease development at chronological age t if a person is in risk group r. (This is the age-specific rate of disease development for a person in risk group r. In what follows, we will suppress the subscript r).

$p(t)$ = hazard rate for clinical surfacing when the disease is t years old – that is, the person has been in the preclinical state for t years. (To simply, we have assumed this rate is not also a function of the age of the person.)

$b(t)$ = prognosis if the disease is t years old when it comes to clinical attention, either as a result of clinical surfacing or detection by a screen. In order to evaluate screening, we would define a benefit measure as some function of prognosis.

$f(t)$ = false-negative rate of the screen when the disease is t years old, i.e., the probability the screen will not detect a tumor that is t years old. Define $f(0)=1$.

In order to simplify these equations, we have not included the possibility of death from causes other then the

disease under consideration. To add this, we would
incorporate in each equation an expression $e^{-D(a)}$, where

$$D(a) = \int_0^a d(u)du$$

and $d(u)$ = age-specific mortality rate from other causes.

We can now write equations to describe the following,
where $P(A)$ = the probability of event A:

    1)  P (disease free at age a) = $e^{-I(a)}$.

    2)  P (preclinical disease at age a)

$$= \int_0^a \left| \frac{i(u)e^{-I(u)}}{} \right| \left| \frac{e^{-P(a-u)}}{} du \right|.$$

The first expression is the probability density function fo
development of the disease at age u. The upper limit of
integration insures that the disease develops prior to age
a, something that must occur if the disease is preclinical
at age a. The lower limit of integration is 0. This
implies that we are calculating probabilities for someone a
the time of birth (current age 0). If we wanted to perform
the calculations for someone of current age $A_c$, all lower
limits of integration that are 0 would be replaced by $A_c$.
The second expression represents the probability that a
disease that is (a - u) years old has not yet clinically
surfaced, something that must be true if disease that
develops at age u is still preclinical at age a.

    3)  P (disease clinically surfaces in some
        interval $\Delta a$ following age a)

$$= \int_0^a i(u)e^{-I(u)} e^{-P(a-u)} \left| \frac{p(a-u)\Delta a}{} \right| du.$$

The expression indicated by the brackets represents the probability someone age a with preclinical disease (a-u) years old will surface in some interval of time Δa after the disease is (a-u) years old.

    4)  Expected prognosis if there is no screening program

$$= \int_0^\infty \int_0^a i(u)e^{-I(u)} e^{-P(a-u)} p(a-u)\Delta a\ b(a-u)\ du\ da.$$

The only addition here is that we assign some prognosis to treatment of the disease when it is (a-u) years old, and we perform the calculation for persons of all ages a. Thus, a is now allowed to vary from $(0, \infty)$. The expression in 4) provides the baseline against which any screening strategy would be evaluated.

To develop the equations when screens are performed, let us consider the case when only one screen is given at age E.

    5)  P (detection at a screen given at age E)

$$= \int_0^E i(u)e^{-I(u)} e^{-P(E-u)} (1-f(E-u))\ du.$$

The expression $(1-f(E-u))$ is the probability the disease, which is E-u years old at the screen, will be detected by the screen.

    6)  P (disease clinically surfaces at age a, a < E) = equation 3 above.

    7)  P (disease clinically surfaces after the screen)

      = P (disease is preclinical at age E and is missed by the screen)

340                                              SHWARTZ AND PLOUGH

+ P (disease initially develops and clinically
  surfaces after the screen)

$$= \int_E^\infty \int_0^E i(u)e^{-I(u)} \, e^{-P(a-u)} \, f(E-u) \, p(a-u)\Delta a \; du \; da$$

$$+ \int_E^\infty \int_E^a i(u)e^{-I(u)} \, e^{-P(a-u)} \, p(a-u)\Delta a \; du \; da.$$

In the first expression, the limits of integration for u
(0, E) insure the disease develops before the screen. The
limits of integration for a (E, ∞) insure the disease
clinically surfaces after the screen. The inclusion of f(E-
u) in the first expression insures only those case missed by
the screen are counted. In the second expression, because
the lower limit of integration for u is E, we insure disease
development does not occur until after the screen.

   8)  Expected prognosis if one screen is given at age E

$$= \int_0^E i(u)e^{-I(u)} \, e^{-P(E-u)} \, b(E-u) \, (1-f(E-u)) \; du \; +$$

$$\int_0^E \int_0^a i(u)e^{-I(u)} \, e^{-P(a-u)} \, p(a-u)\Delta a \; b(a-u) \; du \; da \; +$$

$$\int_E^\infty \int_0^E i(u)e^{-I(u)} \, e^{-P(a-u)} \, f(E-u) \, p(a-u)\Delta a \; b(a-u) \; du \; da +$$

$$\int_E^\infty \int_E^a i(u)e^{-I(u)} \, e^{-P(a-u)} \, p(a-u)\Delta a \; b(a-u) \; du \; da.$$

These are equations 5), 6), and 7) with prognosis
included in the integrand. The improvement in prognosis or
gain realized by performing one screen at age E would be
determined by comparing equation 8 to equation 4.

To summarize, when one screen is being evaluated we
keep track of the following probabilities: the probability
of clinically surfacing before the screen; the probability
of being detected by the screen; the probability of
developing the disease before the screen, being missed by
the screen, and clinically surfacing after the screen; and
the probability of both developing the disease and
clinically surfacing after the screen. When we consider
more than one screen, conceptually we do the same thing.
That is, we determine, for each screen, the probability of
being detected at the screen, and the probability of
clinically surfacing between the given screen and the next
screen. By incorporating a measure of prognosis in the
equations and summing the results for all screens, we can
determine the expected gain in prognosis from any screening
strategy. The only addition to the single screen case is
that some hypotheses must be made about the relationship
between false-negatives on successive screens. Let
$X(v)$ = probability that a tumor will not be detected on $v$
screens. The assumption is usually made that false-
negatives on successive screens are independent. Thus, if
$w(k)$ = age of the tumor on the $k^{th}$ screen,

$$X(v) = \prod_{k=1}^{v} f(w(k)).$$

Assume we want to evaluate a strategy in which screens
are given at ages $E_m$, $m = 1,\ldots,n$. Let $E_0 = 0$ and $E_{n+1} = \infty$.
Then,

9)  P (detection at the screen given at age $E_m$)

$$= \sum_{y=1}^{m} \int_{E_{y-1}}^{E_y} i(u)e^{-I(u)} e^{-P(E_m-u)} X(m-y)(1-f(E_m-u))du.$$

Disease detected at the $\underline{m}^{th}$ screen can develop before the
first screen (between $E_0$ and $E_1$) and be missed by m-1
screens $(X(m-1))$; between the first and second screen
(between $E_1$ and $E_2$) and be missed by m-2 screens $(X(m-2))$;
etc.  The limits of integration and summation allow us to
consider all these possibilities.

10)  P (disease clinically surfaces at age a
       between the $m^{th}$ and $(m+1)^{th}$ screen, i.e.,

$E_m < a < E_{m+1}$,

$$\sum_{y=1}^{m+1} \int_{E_{y-1}}^{\min(E_y, a)} i(u)e^{-I(u)} e^{-P(a-u)} p(a-u)\Delta a \, X(m+1-y)du.$$

Again, the disease can develop before the first screen
(between $E_0$ and $E_1$), be missed by m screens $(X(m))$, and then
surface at a; develop between the first and second screen
(between $E_1$ and $E_2$), be missed by m-1 screens $(X(m-1))$ and
then surface at a; etc.

If we add a measure of prognosis to equations 9) and
10) and sum them over all m, m = 1,...,n, we could determine
the prognosis associated with any screening strategy.  This
could be compared to the baseline prognosis in order to
determine the benefit from screening.

Using this model, a variety of questions of interest
could be examined.  For example,

1) How do the benefits of screening vary as a function of
   the risk group screened? (Vary $i_r(t)$.)

2) How do the benefits of screening vary as a function of
   the age at which screening is started? (Vary $A_c$.)

3) How do the benefits vary as a function of the number of
   screens given and the ages at which the screens are
   given? (Vary n and $E_m$, $m = 1,...,n$.)

4) How do benefits vary as a function of the reliability of
   the screen? (Vary $f(t)$.)

5) Are policy decisions sensitive to the prognostic measure
   used? (Vary how $b(t)$ is measured.)

In addition these equations could easily be extended to
incorporate more than one screening technique, the cost of
false-positives resulting from the screen, and the
possibility of an iatrogenic effect from screening.

3.  Estimation of Model Parameters

    To determine values for the unknown rates, hypotheses
are made about the shape of relevant probability
distributions and parameters estimated so that predictions
from the model correspond to data from clinical studies
(i.e., to "surface" data). For example, in our simple
model, we might hypothesize that the rate of clinical
surfacing $p(t)$ equals p, a constant independent of the age
of the disease. This is equivalent to assuming the
probability density function representing the duration of
preclinical disease follows an exponential distribution.

Let us suppose for this example that at the time of clinical surfacing there is some way to reliably determine the age of the disease, say the number of years that have elapsed since the time of the first malignant cell. (The fact that it is not possible to measure this is the reason why this simple model must be modified for actual application and why formulation and parameterization of deep models is so difficult.) Then, from some large clinical series composed of patients who have not participated in any screening program, patients could be classified by the age of their disease at clinical surfacing. A value of p could then be estimated so that the distribution of people by the age of their disease at the time of clinical surfacing predicted by the model corresponds as closely as possible to the distribution reported from the clinical series.

Once a value of p has been established, i(a) can be estimated. Since the duration of preclinical disease is assumed to follow an exponential distribution, the mean duration of preclinical disease is 1/p. Thus, if we let N(a) = age-specific incidence of clinically surfaced disease at age a, data available from the Third National Cancer Survey, i(a) could be estimated as N(a+1/p), i.e., as the age-specific incidence of clinically surfaced disease 1/p years later.

If we are able to determine the age of the disease at clinical surfacing, then it would be possible to determine from clinical studies yearly mortality or morbidity as a function of the age of the disease at detection. This could provide the basis for formulating a variety of prognostic measures.

The rate of death from other causes could be determined by subtracting the age-specific mortality from the cancer of interest from the overall age-specific mortality rate.

Because there is rarely a definitive means of determining whether or not the disease is present, it is extremely difficult to estimate the false-negative rate of a screening modality. Usually, the analysis is performed under a variety of different assumptions about the false-negative rate of the screen. However, in some cases, the reliability of the screen can be estimated from age-specific prevalence data reported from screening programs.

It should be emphasized that distributional forms and parameter estimates for parts of deep models, e.g., the duration of preclinical disease, can not usually be estimated directly from the data since the age of preclinical disease at clinical surfacing can not usually be measured. Rather, intuitively appealing or mathematically tractable distributions are assumed and parameters are estimated from "surface" data, that is, observed data that result from the unseen process being modelled. For example, the stage of disease at clinical surfacing is often used to estimate parameters for the distribution describing preclinical disease progression. Thus, the relationship of a deep model to data is somewhat more tenuous than usual

statistical models of observable events. For this reason,
model validation becomes particularly important.

## 4. Validation of the Model

There are two different levels of model validation. At
the first level, predictions from the model should
correspond to the data used to estimate parameters of the
model. For example, in our case, the predicted distribution
of patients by age of disease at clinical surfacing should
correspond to data reported from clinical studies. This
type of first order validation insures the initial validity
of the underlying hypotheses.

A second order validation consists of comparing
predictions from the model to data not used in estimating
parameters of the model. For example, again assuming it is
somehow possible to determine the age of disease at the time
of detection, a second order validation in our example might
consist of comparing the distribution of age of the disease
in women detected at a first screen, second screen, and
third screen, and in women who clinically surface between
the different screens predicted by the model to data from an
actual screening program.

Usually, most data available are used in determining
the parameters of the model. Hence, there is often not data
available for second order validation. Nevertheless, second
order validation is important for increasing confidence in
model predications.

## 5. Conclusions

We have outlined a simple "deep" model that could be used to evaluate a cancer screening program. In fact none of the models developed are exactly like the one described above. However, they all incorporate in a variety of different forms the fundamental ideas contained in this model. The main differences concern how the preclinical phase is modelled and the benefit measure determined. Rather than having just one preclinical phase, often a series of disease states are defined such that prognosis is significantly different if the disease is treated in each state. Sometimes the relationship between certain prognostic variables (e.g., tumor size) and time in the preclinical phase is modelled. Disease progression is described by transitions between the disease states or as changes in the prognostic variables. The benefit measure is developed from prognosis associated with each state, or from the status of the prognostic variables at detection reported in clinical studies. A variety of different benefit measures can be considered, though all might be classified into one of three categories: disease related measures - e.g., the probability of detection before invasive disease occurs (cervical cancer) or before lymph node involvement occurs (breast cancer); mortality related measures - e.g., life-expectancy; and economic measures, including often both costs of treatment and costs of premature death.

In conclusion, the models used to evaluate cancer screening programs will consist, in general, of the following:

1) An age-specific rate of disease development;

2) Some definition of the preclinical phase, perhaps in terms of different disease states or the status of prognostic variables;

3) Rates or probability density functions that represent the duration of the preclinical phase, the duration of time in each disease state, or the rate of change in prognostic variables as a function of the time in the preclinical state;

4) Rates or probability density functions representing the time to clinical surfacing;

5) A measure of the reliability of the screen. (Usually, the focus is on the false-negative rate of the screen, in which case one must consider the relationship of false-negatives on successive screens. However, in some cases the false-positive rate is also considered.);

6) Some benefit measure. (This will be determined by the age of the disease at detection, the disease state at detection, or the status of prognostic variables at detection.)

B.  Deep Models Used to Evaluate Cancer Screening Strategies

In the first part of this section, we will briefly discuss several general models that have been used to analyze cancer

screening strategies. We will not discuss more theoretical
formulations that have been proposed, but not applied to the
analysis of any programs (e.g., Voelker, 1976; Thompson and
Disney, 1977). In the second part of this section, we will
discuss the use of these general approaches, plus several
other approaches, to analyze screening strategies for
different site-specific cancers - primarily breast, cervical
and colorectal cancer.

We will confine our discussion to deep models, i.e.,
models based on explicit hypotheses about underlying disease
progression. Thus, for example, certain decision theory
formulations (e.g., Gohagan, *et al.*, 1982) will not be
discussed. This partly reflects our own belief in the
usefulness of deep models to extrapolate from limited
experience and partly the fact that a much more extensive
literature already exists on surface modelling approaches.
However, our focus on deep models is in no way meant to
imply a belief that surface models do not have an important
contribution to make in increasing our understanding of the
benefits of alternative screening strategies.

We will not review literature on the screening tests
available for different cancer sites, but refer the reader
to the American Cancer Society Report on the Cancer-Related
Health Checkup (American Cancer Society, 1980). An
extensive bibliography on cancer screening tests is provided
in this report, as well as a discussion of the basis for the
ACS recommendations on cancer screening.

*1. General Formulations*

Zelen and Feinleib (1969) first proposed the three state model discussed in Section A. They used this model to estimate the average lead time (i.e., how much earlier disease is detected as a result of screening) from a one-time screen. Lead time estimation was extended to multiple screens by Prorok (1976 a, b). Both these formulations were used to estimate the lead time from screening for breast cancer.

Though Galliher has never proposed a general model to evaluate cancer screening strategies, his approach has been used to analyze screening for cervical (Galliher, 1969, 1976, 1977, 1981), breast (Shwartz, 1978 a, b) and colorectal (Thompson and Doyle, 1976; Thompson and Jacobi, 1977) cancer. This approach is summarized in the simplified model discussed in Section A.

Blumenson's formulation (1976) is similar in many respects to the model in Section A. Underlying his approach is a disease model that consists of functions that represent both the probability the disease clinically surfaces and the probability the disease becomes "terminal" (his prognostic measure) as a function of the age of the disease. Screening is evaluated by calculating the probability a person has undetected disease less than some age x, given that the person can come to clinical attention only as a result of screening (i.e., there is no clinical surfacing). For a given screening program, assuming a screen of some reliability, the total number of people in a population who will be detected with disease less than age x can be determined by combining this function with the probability that the disease has not clinically surfaced by age x. When the probability of terminal disease as a function of disease age is considered, a prognostic measure is available for evaluating alternative screening strategies. In a later paper (Blumenson, 1977), this approach was extended to consider the transient situation. This formation has been used to analyze screening strategies for breast cancer based upon a model of disease progression (Bross, *et al.* (1968)).

Dubin (1981) has proposed a multi-stage disease model to evaluate screening programs. In the absence of screening, each stage of the disease is characterized by an incidence function and a survival function. In order to determine the benefits of screening, a screening "sensitivity" function is defined. This function describes the proportion of disease that, in the absence of screening, would have been diagnosed in some stage at time t after screening that is diagnosed in an earlier stage as a result of screening. Using this function, formulas are developed for the incidence of disease at each stage that results when screening is performed. Finally, a survival function is defined for each disease stage when the stage is detected by screening. From the incidence and survival functions with and without screening, a variety of benefit measures can be calculated.

This model, which has been used to estimate the long-term benefits of an initial screen in the HIP breast cancer screening program (Dubin, 1979), is in some sense at the border between surface models and deep models. Though no explicit hypotheses are made about the rate of disease progression, the screening "sensitivity" function incorporates implicitly hypotheses both about the rate of disease progression and screen reliability.

Albert *et al.* have developed a model to analyze population screening strategies. A disease is represented by a series of states $S_0$, $S_1$,...,$S_n$, $Y$, where $S_0$ is the healthy state, $Y$ is clinical disease and $S_i$, $i=1$,...,$n$ are preclinical disease states used to measure disease progression. Associated with each person is a vector ($X_0$, $X_1$,...,$X_n$, $Y$, $A$), where $X_0$ = age of the person at entry into state 1, $X_i$ = duration of time in state i, $i=1$,...,$n$, $Y$ = duration of time to clinical surfacing from $X_0$, and $A$ = current age of the person. If $X_0 = \infty$, a person will never develop the disease; if

$$\sum_{i=1}^{k} X_i > Y,$$

the person will surface with the disease at or before state k. With the exception of $A$, this vector is fixed for each person. In a population of people there are a variety of ($X_0$,...,$X_n$, $Y$, $A$) values. Thus, a joint density of ($X_0$,...,$X_n$, $Y$, $A$) at any time $t$ can be defined over a population. This density function provides the basis for

the analysis. From this density function, a formal
definition is given to traditional concepts used in
epidemiology (Albert *et al.*, 1978 a). Also, the density
function is used to derive the evolution of the natural
history of cancer in a population subject to screening and
natural demographic forces (Albert *et al.*, 1978 b). That
is, the authors have shown how the disease status of the
population at time a+t can be determined as a function of
the immigration rate and the disease status of those who
immigrate, the probability of being screened as a function
of disease status, the probability of a positive screen as a
function of disease status, and the rate of death from other
causes. By associating a benefit measure with disease
status, the benefit of a screening strategy can be
determined. Perhaps most interesting, this formulation has
been used to develop maximum likelihood estimates of the
joint distribution of holding times in the disease states
(Louis *et al.*, 1978). The model has been used to analyze
screening strategies for cervical cancer.

Knox (1973) has developed a simulation model to
evaluate screening strategies for different diseases. The
model consists of a specification of a population to be
screened, especially its age and sex distribution; the
natural history of the disease process, i.e., a series of
disease states and a specification of the probability of
making transitions between the states, where the
probabilities may vary as a function of the age of the

person and the length of time already in the state; the
false-negative rate, false-positive rate, and acceptability
of the screening procedure; the effect of diagnosis and
subsequent treatment upon the natural history of the
disease; and the screening policy to be evaluated. Needed
specifications of natural history and screen reliability are
estimated by an iterative procedure that consists of making
proposals about, for example, possible transition rates
between disease states, computing results, comparing results
with available observations and readjusting the
specifications. In general, it has not been possible with
existing data to distinguish between different natural
history hypotheses. Hence, analysis is usually performed
under different assumptions about natural history and screen
reliability. This approach has been used to evaluate
screening strategies for cervical and breast cancer.

A group at Erasmus University, Rotterdam has been
working on extensions of Knox's model, using "micro-level"
Monte-Carlo simulation (van Oortmarssen *et al.* 1980). They
have applied their approach to the analysis of cervical
cancer screening in the Netherlands and plan applications to
screening for other cancers.

The formulation most different from those already
discussed is that proposed by Eddy (1980). Whereas most
approaches model the disease process and then evaluate
screening by superimposing on the disease process a strategy
with a screen of some reliability, Eddy models the

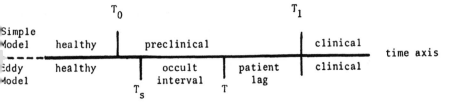

Figure 1    Eddy Model Compared to the Simple Model


interaction between the screen and the disease.    To
illustrate this approach consider Figure 1.  $T_0$ represents
the time of disease development and $T_1$ the time of clinical
surfacing.    Define a point T as a reference point.    The way
this point is defined will differ for the different types of
cancer.    For example, in breast cancer it is the time
disease can first be detected by a physical examination.
There is some other point $T_s$ at which the disease is first
detectable by a screen (e.g., mammography).    $T_s$, T and $T_1$
can occur in any order.    The important assumption made by
Eddy is that once the disease is detectable by a screening
modality (e.g., once $T_s$ occurs if the modality is
mammography or T occurs if the modality is physical
examination) any screen using that modality given after the
point will always detect the disease.    This assumption
replaces the assumption usually made in other analyses that
successive screens are independent.    In our original
formulation, $T_0$ is the time the disease is first
theoretically detectable in an individual.    In the Eddy
formulation, $T_s$ is the time the disease is actually

detectable by mammography in an individual. Thus, $T_s$ will occur after $T_0$. The interval $(T-T_s)$ is called the mammogram (for the case of breast cancer) interval and $(T_1-T)$ the patient interval. The duration of the mammogram interval and the patient interval are described by probability distributions.

Using this formulation, it is possible to write formulas for the probability, given some screening program, that the disease will be detected at a screen by different screening modalities and that a person surfaces clinically between screens. Parameters for the probability distributions representing the mammogram interval and patient interval are determined so that model predictions correspond to the data. In addition, formulas are developed that describe the age of the disease as a function of how it was detected. From these formulas, a function that describes mortality (the mortality function) is developed so that mortality by mode of detection predicted by the model corresponds to data from screening programs. Also, estimates are made of the age-specific rate of disease development, the probability of a false-positive test result, and relevant dollar costs. Given these estimates, alternative screening strategies can be evaluated using both economic and mortality type measures.

2.  *Site Specific Analysis*

a.  *Breast cancer*.  It was demonstrated in a randomized controlled trial at the Health Insurance Plan of Greater New

York (HIP) that when annual screens with mammography and
clinical examination were offered to a population of women,
mortality from breast cancer was reduced.  The success of
this experiment motivated the American Cancer Society and
the National Cancer Institute to jointly establish twenty-
seven Breast Cancer Detection Demonstration Projects (BCDDP)
around the country.  However, both the HIP trial and the
BCDDP have raised questions about breast cancer screening
strategies.  Some of the more important of these questions
are the following:

1)  From the HIP trial, we know only the value of offering
    annual screening for five years to a population of
    women.  This population was composed of women of
    different ages in different risk groups who had varying
    degrees of acceptance of the offered annual screens
    (i.e., some never appeared for the screen, some once,
    some twice, etc.).  However, to actually recommend a
    particular screening strategy to a cohort of women of
    some age in some risk group, we would like to know the
    benefits that would be realized by the cohort if they
    complied with the recommendation.  What would these
    benefits be?

2)  The HIP trial evaluated the benefits of offering
    clinical examination and mammography yearly.  To
    actually plan a screening program, one would be
    interested in the benefits of offering mammography and a
    clinical examination every two years, offering a

clinical examination yearly and mammography every two
years, etc. What are these benefits?

3) To what extent do radiation risks from mammography
offset the benefits from earlier detection?

4) The HIP trial did not show any benefit from screening
for women under age 50. This may be due to two factors:
one, mammography used by HIP was less reliable in
younger women; two, disease in younger women is
different than disease in older women such that there is
less value in earlier detection in younger women. There
are indications from the BCDDP that the reliability of
mammography in younger women has significantly improved.
Thus, what priority that should be assigned to a new
randomized controlled trial to evaluate the use of
mammography to screen women age 40 to 50?

In what follows, we discuss several mathematical models
that are able to consider these types of questions.

Shwartz (1978 a, b) has developed a model to analyze
screening strategies for breast cancer that is similar to
the "simple" model described in Section A. In his model,
disease progression is defined in terms of the status of two
prognostic variables - tumor size and number of axillary
lymph nodes involved. Specifically, it is hypothesized that
(a) tumors grow exponentially at a rate selected from a
probability distribution determined from available data on
tumor doubling times; (b) the rate of lymph node involvement
at tumor age t is a linear function of the size of the tumor

at age t and the growth rate of the tumor; and (c) the rate
of clinical surfacing at tumor age t is a linear function of
the size of the tumor at age t and its growth rate. The
parameters for each of the linear functions are estimated so
that the tumor size and lymph node status of patients who
surface clinically with the disease correspond to available
data. The age-specific rate of disease development is
estimated so that, given the rate of clinical surfacing
already determined, the age-specific incidence of the
disease predicted by the model corresponds to available data
from the Third National Cancer Survey. Benefit measures are
determined from data on the five-year survival rate and
five-year rate of disease recurrence of patients as a
function of tumor size and number of lymph nodes involved.
The false-negative rate of mammography and physical
examination are determined from overall prevalence at the
initial screen given by HIP and from the percentage of
tumors detected on physical examination alone reported both
from HIP and BCDDP. Data that were not used in parameter
estimation from the two screening programs were used for
second order verification. Both the HIP program and the
BCDDP were simulated, incorporating the age distribution of
women and adherence to the screening program reported by
HIP, and the age distribution and risk levels reported from
the BCDDP. Under a variety of different assumptions, the
results from the simulation were, in general, in accord with
reported data (Shwartz, 1982).

The analysis of this model has been performed under a
variety of alternative assumptions about the rate of disease
progression, long-term prognosis following detection, the
reliability of the screen, and the age and risk category of
women screened. Among the important conclusions from the
analysis are the following:

1) Yearly screening with mammography and a physical
   examination will result in about 40% of the gain in
   life-expectancy that would be realized if mortality
   from breast cancer were eliminated.

2) If screening were performed every two years rather
   than yearly, about 30% of the impact of an annual
   program would be lost.

3) If screening were performed only with a physical
   examination, about 30% of the impact of an annual
   program with the combined modality would be lost.

4) The addition of mammography to yearly screening
   with a physical examination may cause as a result
   of the radiation from mammography as many cases of
   breast cancer as resulting lives saved. However,
   if benefits are measured in terms of increased
   life-expectancy, the radiation risk from
   mammography is minimal. This is because most of
   the induced cases of breast cancer are detected
   sufficiently early in their disease course and
   sufficiently late in life that they do not result
   in premature death.

5) If annual screening is started at age 50 rather than age 40, almost 40% of the possible benefits in terms of improved life-expectancy from annual screening starting at age 40 would be lost; if the benefit measure is death from breast cancer (a measure not sensitive to age at death), only 18% of the potential benefit would be lost. (It is important to emphasize that this result is based on the assumption that the lack of benefits for younger women in the HIP trial was because of the unreliability of the screen and not differences in the disease process in younger women versus older women.)

6) Self examinations regularly performed have the potential to significantly reduce the mortality threat of breast cancer.

In addition, this model has been used to estimate lead time and length bias in a breast cancer screening program (Shwartz, 1980).

The Blumenson-Bross model (Bross et al., 1968; Blumenson and Bross, 1969), which preceded the Shwartz model, is conceptually similar to the Shwartz model. In this model, hypotheses are made about the relationship between the duration of preclinical disease and three disease related variables: tumor size, number of axillary lymph nodes involved and occurrence of distant metastasis. Specifically, it is assumed that tumor growth is

exponential, the rate of lymph node involvement is constant,
and the rate of distant metastasis is a linear function of
time. Thus, in the initial formulation there are three
unknown parameters - the rate of tumor growth, the rate of
lymph node involvement, and a parameter defining the
relationship between the rate of distant metastasis and
time. The unknown parameters are estimated so that
predictions from the model correspond to data on patients
who have clinically surfaced with breast cancer. It was
found that to fit the data, it was necessary to postulate
two types of breast cancer, each of which had different
values for the three unknown parameters. In one type of
breast cancer, nodes become involved and metastasis occurs
very early; in the other type, nodal involvement and
metastasis rarely occur. Thus, this is a fairly
biologically "predetermined" model of the disease, i.e., one
in which prognosis depends more on characteristics inherent
at the time of disease development than on the duration of
preclinical disease. In contrast, the Shwartz model, a
multi-disease model, postulates the disease process is much
less biologically predetermined, i.e., prognosis is much
more sensitive to the age of the disease at detection. This
difference results from differences in the form of the
hypothesized rates in the two models.

Blumenson has used this model in conjunction with his
framework for analyzing alternative screening strategies to
evaluate screening for breast cancer. Probably due to the

extent of biological "predeterminism" incorporated in his disease model, the predicted potential gains from screening are less than indicated by the HIP screening program. Thus, the criteria for second order verification are not met. It should be noted that this does not necessarily mean the model is not useful. Specifically, this does not negate its usefulness for analyzing the relative gains of alternative strategies, nor for considering the sensitivity of findings to alternative formulations. Among the most important conclusions from Blumenson's analyses are the following:

1) If screening with mammography is performed every other year rather than yearly, slightly over 40% of the potential gains from yearly screening will be lost (versus 30% estimated by Shwartz).

2) If screening is started at age 50 rather than age 40, slightly over 20% of the potential gains will be lost (versus slightly under 20% estimated by Shwartz when a similar benefit measure is used).

3) Use of a very unreliable first level screen, followed by mammography if there is a positive result, is a much more cost-effective strategy than only using mammography (Blumenson, 1977).

Eddy (1980), in the context of his theory of screening, has analyzed screening strategies for breast cancer. The parameters of the probability distributions representing the mammogram interval and patient interval are estimated so that the proportion of women detected at each examination by the different modalities (i.e., mammography and physical examination) and the proportion surfacing between different screens predicted by the model correspond to data reported by HIP. The mortality functions are estimated from the survival experiences of these different groups of women, as reported by HIP. The rate of disease development is determined from incidence data reported by the Third National Cancer Survey. In addition, Eddy has incorporated false-positive rates of the respective tests in his analysis. Eddy has performed probably the most extensive analysis of different screening scenarios. Though Eddy performed the analysis for a 50 year old woman at normal risk and Shwartz for a 40 year old woman at normal risk, in general their conclusions are similar (Eddy and Shwartz, 1982).

In his model, Knox (1975) postulates the following disease states: preclinical (i.e., non-self-detectable), early clinical (potentially self-detectable), and late clinical. At the time of tumor development, the tumor is assumed to be in one of two grades: high grade or low grade. It is assumed one half of the tumors are of each type. These grades are not meant to imply histologic grades, but are modelling constructs used to reflect "better" tumors and "worse" tumors. For each type of tumor, the age-specific rate of disease development, progression rates between states, and mortality rates as a function of

disease state were estimated and adjusted iteratively until
the output produced by the model corresponded to available
data on incidence, prevalence and mortality. The false-
positive and false-negative rates for the joint mammography-
clinical examination modality were determined so that
predictions from the model replicated HIP data.

Among the conclusions from Knox's model are that annual
screening with a physical examination and mammography will
result in a 25% reduction in mortality from breast cancer
(compared to 40% estimated by Shwartz). Almost 80% of this
gain would be realized if only a physical examination were
performed (compared to 70% estimated by Shwartz). Knox,
similar to Blumenson, determines there are potential
benefits from a first level, very unreliable, screen to
detect breast cancer.

Kirch and Klein (1978) have developed two models of the
disease that differ only in the size of the tumor at disease
development. In one model, which represents tumors that
will be detected by mammography, disease development occurs
when a tumor is 0.5 cm in diameter; in the other model,
which represents tumors detected by a physical examination,
disease development occurs when the tumor is 1 cm in
diameter. After disease development, tumor growth is
assumed to be exponential. The distribution of the rate of
tumor growth, determined from data on tumor doubling times,
is assumed lognormal. It is further assumed that in the
absence of any screening a woman will examine herself

"diligently at monthly intervals and be able to reliably
detect a 1.6 cm tumor". Because of the distribution of
tumor growth rates, for each model a distribution is induced
on time to clinical surfacing. By superimposing a screening
strategy on each model, it is possible to determine both the
probability of being detected at a screen and the expected
number of tumor doublings until detection. By relating the
expected number of tumor doublings to the probability of
lymph node involvement (using data on clinically surfaced
cases of breast cancer) the expected probability of lymph
node involvement for different strategies can be determined.
This is the benefit measure used in the Kirch-Klein model.

Because of the assumption that in the absence of
screening tumors will be detected at 1.6 cm by self-
examinations (plus some additional "presentation delay") and
the manner in which the relationship between the probability
of lymph node involvement and tumor doublings is determined,
this model tends to accord less potential to screening than
the Shwartz model. For example, Kirch and Klein estimate
that in the absence of screening, 42% percent of patients
will have nodal involvement at treatment. Yearly screening
with mammography and a physical examination will reduce the
percentage with nodal involvement to 34%. Shwartz estimates
these two percentages as 49% and 22% respectively.
Similarly to Shwartz, Kirch and Klein estimate potential
benefits from regular self-examinations.

In regard to some of the questions posed above, we
might draw the following conclusions from these modelling
efforts:

1)  There is no indication annual screening with
    mammography and a clinical examination will come
    close to eliminating the mortality threat from
    breast cancer.

2)  Policy decisions are sensitive to the benefit
    measure used.  For example, if one's criteria is to
    maximize life-expectancy, the radiation risks from
    mammography are minimal; if one's criteria is to
    compare cases of breast cancer to resulting lives
    saved by the addition of mammography to an annual
    clinical examination, the marginal value of
    mammography becomes more  questionable.

3)  If screening under age fifty was not of value in
    the HIP trial primarily because of the technology
    used by HIP (technology which results from the
    BCDDP indicate has improved), there may be great
    potential to be realized in terms of improved life-
    expectancy by screening women age forty to fifty.
    This would suggest some priority be assigned to a
    trial to determine whether screening with current
    technology can improve mortality in younger women.

4)  Annual clinical examination alone can realize a
    significant percentage of the gain realized by

screening with the combined modality.  Since
clinical examinations are significantly cheaper
than mammography, this might suggest a high
priority be assigned to a trial to evaluate
clinical examinations.

5)  Breast self-examinations may have significant
potential to reduce mortality from breast cancer.
Though the analyses of the benefits of breast self-
examinations are particularly unreliable due to
lack of data and due to formulations that may tend
to over-emphasize benefits, its seems this
potential should be further explored.

6)  It is important to consider the possibility of
sequential screening strategies for the early
detection of breast cancer.  Thus, the fact that
some potential screening technique is unreliable
does not obviate it usefulness as a component in a
sequential screening strategy.

b.  *Cervical cancer*.  The use of the Pap smear as a
screening technique to detect cervical cancer is widespread.
However, no randomized controlled trials have been performed
to evaluate the Pap smear.  Hence, though there is
circumstantial evidence indicating the Pap smear has reduced
mortality from cervical cancer, the efficacy of this
technique is still somewhat controversial.  The controversy
centers around two points:  one, the extent to which and
rate at which carcinoma in situ (c.i.s.), the disease

usually detected by the Pap smear, would, if untreated, progress to invasive cervical cancer; and two, the fact that the decline in mortality that has occurred concurrent with the Pap smear began before widespread use of the Pap smear to screen for cervical cancer. Nevertheless, despite the controversy, belief in the efficacy of the Pap smear is substantial enough and its use widespread enough that a randomized controlled trial is both ethically questionable and practically infeasible.

Until recently, the usual recommendation has been that women should have a yearly Pap smear. However, there is little information on the relative benefits from alternative screening frequencies, particularly as a function of a person's risk category for cervical cancer. This is the question on which analyses of screening strategies for cervical cancer have focused - that is, what are the relative benefits of alternative screening frequencies for women in different risk categories. These analyses have been performed under different assumptions about the reliability of the Pap smear, the extent to which carcinoma in situ will progress to invasive disease, and the duration of carcinoma in situ (an important determinant of the frequency at which screens should be performed).

Lincoln and Weiss (1964) were the first to propose a model of cervical cancer (or any cancer) as the basis for analyzing serial screening. Their intent was to present a "plausible model" rather than one closely correlated with specific data. Their model consists of the following:

1) a hypothesized rate of disease development;

2) an assumption that disease progression is
   independent of the age of the patient;

3) a false-negative rate of the screen, which in some
   cases varies as a function of the age of the tumor;

4) an assumption that the disease can be detected only
   by the screen examination.

This last assumption relieves them of the necessity of
hypothesizing a rate of clinical surfacing.  In the
analysis, the expected age of the tumor as a function of
different screening frequencies and different assumptions
about the false-negative rate of the screen is determined.
Among their conclusions is the importance of performing Pap
smears at regular planned intervals as opposed to random
intervals.

Galliher, in his most recent work (1981), has proposed
a. model of cervical cancer with two strains of early
neoplasia:  a "progressive" strain that would advance to
invasive cancer if not detected and a "non-progressive"
strain that would not advance.  To determine the age-
specific rate of onset of the early phase of each strain,
Galliher uses the age distribution of incidence rates from
the British Columbia cervical cancer screening program and
assumes that the age-specific rate for the non-progressive
strain is 54% of that for the progressive strain (based on
the fact that the total incidence rate of diagnoses of
dysplasia was approximately 54% of that for carcinoma in

situ). The duration of the early phase of the progressive
strain (e.g., carcinoma in situ) is described by an
exponential distribution with a mean of 10 years. Duration
is assumed not to depend on the age of disease onset. Once
the disease progresses to an invasive phase, two concurrent
processes are initiated: one, a tendency for the disease to
become fatal; two, a tendency for the disease to clinically
surface. Thus, even in the absence of screening, some cases
of the disease would clinically surface before becoming
fatal and a cure would be possible; in other cases, the
disease would become fatal before clinical surfacing. Both
the duration of time to fatal disease and the duration of
time to clinical surfacing are modelled by exponential
probability distributions. The parameters of these two
distributions are determined from published data on survival
of cancer patients during the pre-Pap smear period. (From
these data, it was determined that 54% of the patients had
fatal disease at clinical surfacing. The two parameters,
adjusted to replicate this percentage, are estimated as 3.5
years for the average time to clinical surfacing and 4 years
for the average time to conversion to fatal disease.) From
published data, costs are determined for treatment in the
various disease states and an economic penalty assigned for
premature death.

Under various assumptions about the false-negative rate
of the screen, cost, the age and risk category of the woman
under consideration, and the disease process, Galliher

calculates the optimal frequency of the Pap smear so total
economic costs are minimized. In general, he concludes that
if one's objective is to minimize total economic loss (i.e.,
including economic loss from premature death), the optimal
strategy for a woman at normal risk is 22 Pap smears, done
about every 15 months from age 22 to 35, the interval
lengthening gradually thereafter.

Coppleson and Brown (1975, 1976) have developed a model
of cervical cancer consisting of the states healthy,
dysplasia, carcinoma in situ and invasive disease. Similar
to Galliher, the assumption is made that the duration of
time in each state can be described by an exponential
probability distribution. The parameters for the
exponential distributions are determined from published data
on cervical cancer screening, adjusted to account for an
estimated false-negative rate of the Pap smear of about 25%.
The authors determine it is not possible to replicate the
data unless it is assumed that the rates of progression of
the disease are dependent on the age of the person and that,
in the early years, carcinoma in situ may regress to
dysplasia or healthy.

Screening strategies are investigated using the
following benefit measures: the number of cases of invasive
cancer prevented, the expected increased years of life that
result from preventing a case of invasive cancer (this is
calculated under the assumption that all cases of invasive
cancer will die), and a weighted expected years of life

realized by preventing a case of cervical cancer. In
general, they determine that the percentage of possible
savings is not sensitive to the benefit measure used, and
that with 10 screens, 50% of the maximum potential savings
could be realized and with 20 screens, 75%.

Eddy (1977) has estimated that the mean duration of
c.i.s. is about 5 years, though some disease may progress
quite rapidly (with a mean duration of 1 year). Based on
estimates of physicians, the average patient delay from
onset of symptoms until clinical surfacing is assumed to be
1.5 years. Mortality functions are estimated from survival
data from screening programs. Incorporating these estimates
in his overall theory of screening, and assuming that about
40% of carcinoma in situ will regress, Eddy calculates that
over 85% of the increase in life-expectancy for a woman at
normal risk that can be realized by yearly screening would
be realized if screening were performed every 4 years.

Knox (1973) has used his simulation model to evaluate
screening strategies for cervical cancer. A large number of
disease states are considered, e.g., dysplasia, regressive
type; dysplasia, progressive type; carcinoma in situ,
younger type; carcinomas in situ, older type; clinical
invasive disease, early; clinical invasive disease, late;
coned normal; coned occult invasive; hysterectomy not for
carcinoma, etc. The probabilities of transferring to
different states in each time period "were developed
iteratively by making a series of proposals about transfer

rates between pathological types, computing results,
comparing the results with available observations, and
readjusting the specifications." The basic observations
used for this fitting process are mainly age-specific
prevalence data and mortality and onset data. From the
available data, it was not possible to distinguish between
several possible natural histories. Hence, the analysis is
performed under two assumptions about the disease process,
differing mainly in the extent to which carcinoma in situ
progresses to invasive disease. The analysis is also
performed under different assumptions about the false-
negative rate of the screen. In general, it is determined
that under assumptions about the disease that accord maximum
potential to the Pap smear, almost 90% of deaths from
cervical cancer could be saved by 10 tests; under more
pessimistic assumptions, slightly more than 60% of the
deaths could be eliminated.

van Oortmarssen (1979) has analyzed a model of cervical
cancer with disease states dysplasia, carcinoma in situ and
four stages of invasive disease. Two types of disease are
assumed, a slow progressing disease which occurs in women
under age 35 and a fast progressing disease which occurs in
women over age 35. Spontaneous regression is not allowed.
Duration in each disease state is described by an
exponential distribution. Dysplasia, "a state whose
detection does not alter the possibilities of developing
cancer" and thus more "a state of (very) high risk than a

true predecessor of cancer", is assumed to have an average duration of 12 years. For slow disease, the other states have average durations (in years) of 20, 3.5, 2.3, 2; for the fast disease, durations are 3, 2, 2, 2, 1. Given assumptions about survival following detection in each state, the sensitivity of the Pap smear, and compliance with screening recommendations, the long term benefits of two screening policies are calculated:  the current policy in The Netherlands of giving screens every 3 years between age 35 and 53 and a policy where screens are given at age 40, 45, and 50.  In Habbema *et al.* (1980), the basis for assumptions about duration times are further discussed and simulated results compared to actual data from The Netherlands.

Albert *et al.* (1978 a, b) have used their model to compare a policy that screens all women over 30 annually to several "threshold adaptive" policies, that is, policies in which only women in that age stratum that will benefit most from screening in a given year (a group determined in their model) are screened in that year.  In general, they estimate that by using "threshold adaptive" policies instead of annually screening all women, the same benefit can be realized at 60% of the effort.

Gohagan and Swift (1981) use a decision theory framework to analyze both lifetime screening schedules and personal decision-making for individual women in real time. Using a decision tree approach, an expression is derived for

a threshold level such that if the probability a woman has
carcinoma in situ or more advanced disease is greater than
this threshold, a smear should be done.  The threshold level
is a function of the cost of the smear, the cost of biopsy,
the expected cost of treatment, the expected cost of a
missed cancer (which depends on the rate of disease
progression) and the false-positive and false-negative rate
of the Pap smear.  To determine the expected cost of a
missed cancer, an exponential probability model of disease
progression which includes the stages dysplasia, carcinoma
in situ and invasive disease is fitted to published data.
Results were insensitive to variations in the average
duration of carcinoma in situ from 8 to 15 years.  The
probability of disease is estimated from age-specific
incidence data modified to reflect risk factors.  It is
concluded that except in cases of exceptionally high risk,
triennial to pentennial screening is most appropriate for
asymptomatic women after age 23.  Annual screening would be
appropriate only for women having 10 times average risk or
more.  This level of risk would be unlikely unless Herpes II
antibodies were detected or dysplasia previously diagnosed.

To summarize, the main policy conclusion from the
various analyses of cervical cancer screening seems to be
that screening women at normal risk less often than yearly
may be cost-effective.

c.  *Colo-rectal cancer*.  Until recently, proctosigmoidoscopy
has been the principal modality used to screen for colo-

rectal cancer. Though there have been no randomized
controlled trials and little data are available on the
relative value of different screening frequencies, the
recommendation has often been made that all persons over age
40 have a proctosigmiodoscopy at least every three years.
However, because proctosigmiodoscopy is an unpleasant and
somewhat painful procedure, this recommendation is less
widely followed than, for example, the recommendation that a
woman receive an annual Pap smear.

The potential value of proctosigmiodoscopy lies in its
ability to detect adenomatous polyps in the distal 25 cm. of
the large bowel. Controversy surrounds estimates of the
percentage of these polyps that would progress to invasive
disease and, more importantly, the percentage of invasive
disease that arises from the polyps (versus _de novo_ from the
mucosa). To the extent most invasive disease arises from
polyps, identification and removal of polyps by
proctosigmoidoscopy will be of value in reducing the threat
of colorectal cancer.

The occult blood test, which has the potential to
detect early invasive disease throughout the colon, is
currently being evaluated by a randomized controlled trial.
Though there is as yet little evidence screening with the
occult blood test will reduce mortality from colorectal
cancer, and though there is concern about the potential high
cost of false-positives, this test does appear to have
potential as a screening modality for colorectal cancer.

Thompson and Jacobi (1977) have developed a model to
analyze screening with both proctosigmoidoscopy and an
occult blood test. Six disease states are defined: healthy,
a precancerous state (corresponding to adenomatous polyps),
and four invasive states (corresponding to the four
categories in Duke's classification – a pathological staging
system that measures the extent of disease by the categories
A, B, C, D). The age-specific rate of development of
adenomas is calculated from data on the age-specific
prevalence of adenomas at autopsy. From very limited
published data, the authors were able to estimate the
probability that an adenomatous polyp becomes invasive
disease as a function of the duration of time since
development of the polyp. Once a person enters an invasive
disease state (e.g., a person enters a Duke's A state) two
processes are initiated: one, a tendency to progress to the
next invasive disease state (e.g., Duke's B); and, two, a
tendency to clinically surface in that disease state. The
assumption was made that each of these tendencies could be
represented by a constant (hazard) rate (equivalent to an
assumption of exponential duration in each state). The two
hazard rates associated with each disease state were
constrained by the age-specific incidence of clinically
surfaced disease in each disease state. Within these
constraints, the rates were arbitrarily chosen. Part of the
analysis consisted of determining the sensitivity of the
results to variations in these transition rates. A benefit

measure was developed by associating both prognosis and economic costs with the state of the disease at treatment. Analysis was performed under different assumptions about the benefit criterion, the role of adenomas as precursors to invasive disease, the accuracy of the screening technique and the length of time the cancer remains in an early stage.

Under the assumption that all invasive disease arises from polyps, it was determined that a significant fraction of the additional life-expectancy that could be realized if proctosigmoidoscopy were performed yearly, would be realized if it was performed every four to five years. If all invasive disease arose de novo, it would be beneficial to screen somewhat more frequently, but with ten screens given every four years, there is only about 10% difference in possible additional life-expectancy realized. An analysis of occult blood tests led the authors to conclude "that the maintenances of a high level of protection against colorectal cancer may require frequent screening (e.g., annually)."

Eddy (1980) has analyzed screening strategies for colorectal cancer. The following assumptions, determined from his analysis of available data, were incorporated:

1)  5% of adenomatous polyps will progress to invasive disease.

2)  75% of invasive disease arises from polyps (the analysis was also performed under the assumption

that only 25% of invasive disease arose from
polyps).

3)  the mean duration of the occult interval
    (corresponding to the mammogram interval for breast
    cancer) for invasive disease developing from polyps
    is 7 years; for lesions arising de novo, the mean
    duration is 1 year.

4)  the mean duration of the patient interval for the
    occult blood test is 1.3 years.

5)  the false-positive rate of the occult blood test is
    1% (analysis were also performed at values of 5%
    and 10%).

In addition, mortality rates were estimated from survival
data by stage of disease at detection and from state of
disease by mode of detection.

    In general, Eddy's conclusion are similar to those of
Thompson and Jacobi.  That is, almost all the benefit from
performing proctosigmoidoscopy and an occult blood test
yearly will be realized if the occult blood test is
performed yearly and proctosigmoidoscopy every 4 to 5 years.
Eddy also has confirmed the importance of keeping the false-
positive rate associated with the occult blood test low if
the dollar costs from a screening program are to be within
reasonable bounds.

d.  Other cancers.  The chest x-ray and sputum cytology are
the two screening tests sufficiently well developed for

consideration as modalities to screen for lung cancer.
Controlled trials are being undertaken to evaluate these
modalities.  So far, there are no indications that screening
with these techniques will reduce mortality from lung cancer
and, at this point,one must be pessimistic about their
potential value.  Urinalysis  and urine cytology are two
potential screening techniques for bladder cancer.  There is
little information available on the efficacy of these
techniques.

Eddy (1980) had undertaken analyses of screening for
both lung and bladder cancer and determined, based on both
current evidence and optimistic assumptions about unknown
parameters, that screening for neither of these disease is
currently warranted.

Ellwein (1978, 1979 a, 1979 b) has developed a model of
bladder cancer to evaluate research, screening, diagnostic
and treatment strategies.  Progression of disease from
normal tissue through precursor lesions, superficial lesions
and locally invasive disease to eventual metastatic disease
is described.  This model is currently being used by the
National Bladder Cancer Project as an aid in making resource
allocation decisions.

3.  *Conclusions*

In this section, we have presented the types of "deep"
models that have been used to analyze cancer screening
strategies in the general population.  These models have the

potential to provide information useful to aid in the types
of both policy and strategic decisions that must be made if
screening is to be a major component of cancer control
activities.  They are useful because they provide an
explicit basis for extrapolating from experimental evidence
in order to examine tradeoffs necessitated by different
decision options.  Further, they can often be of aid in
clarifying debate surrounding alternatives by demonstrating
how conclusions are sensitive to different assumptions.

There is indication that several of these models have
been considered in actual decision situations.  Shwartz's
model of breast cancer has provided some guidance in the
development of the breast cancer screening component of the
Metropolitan Detroit Cancer Control Program (Swanson and
Shaw, 1978).  Eddy (1980) has a most interesting discussion
of the use of his analysis by the Blue Cross/Blue Shield
Association to aid in designing an insurance benefit to
cover the costs of screening for cancer.  Also, Eddy's
analyses were used by the American Cancer Society in
developing their new recommendations on cancer screening
(American Cancer Society, 1980).  However, in general the
analyses have not been performed to consider the design of
particular screening programs.  Rather, they have been
performed to analyze and clarify issues at the policy level.
As such, it is difficult to determine their contribution
relative to other forces that affect policy.

In conclusion it should be emphasized that models such as these can be aids to a decision-maker. However, the issues involved in designing and implementing screening programs are complex. Further, assumptions and implications are made in models that are often difficult to verify. Hence, one should not uncritically take at face value model results, but use model results as one input into a decision process that may include a host of other factors.

III. SCREENING STRATEGIES FOR THE OCCUPATIONAL CANCER PROBLEM

The models discussed in the previous section were developed to consider screening strategies for cancer in the general population. In this section, we will discuss models to aid in planning screening strategies in industrial populations. The main thesis we will develop is the following: due to the different type of data available about cancer in industrial populations, the focus of control strategies, and hence modelling efforts, should be different. Specifically, whereas in the general population emphasis is on earlier disease detection, in industrial populations more benefits seem possible by identifying potential carcinogens and limiting worker exposure. The focus of models to support the screening of substances will be somewhat different than that to support screening of people.

In what follows, we will review the way in which knowledge about cancer in the general population differs from cancer in industrial populations, discuss methods to identify carcinogenic substances in the workplace, and then outline a modelling approach to support the primary prevention activities possible by screening substances for carcinogenicity.

The focus of this part of the chapter will be different than Part II. Attempts to develop models to evaluate cancer screening programs in the general population have been ongoing for over 15 years. In Part II, we attempted to provide an integrating framework for these efforts and then to review individual models in the context of this framework. However, there has been almost no work on the development of cost-effective strategies to screen substances for carcinogenicity, due primarily to lack of data on the characteristics of the screening tests. A rapidly expanding data base will soon offer the potential to analyze alternative strategies to screen substance. Thus, our focus in Part III is to justify the emphasis on screening substances, to suggest a conceptual framework for analyzing screening strategies of substances, and to illustrate this framework with an example.

A.  General Population Cancers

The cancers for which general population screening is either currently being undertaken or contemplated can be characterized as cancers of relatively high incidence. As a

result, it has been possible to collect a large body of
information on treated disease and some information on
disease progression. From these extensive clinically-based
treatment data, two important characteristics of these
cancers are evident: (1) it is possible to classify the
disease into more or less distinct stages and (2) treatment
of the disease at an earlier rather than later stage results
in more beneficial mortality rates.

It is important to emphasize that this evidence in and
of itself does not indicate the efficacy of early detection.
For, "early" and "late" do not usually refer to points on a
time scale, but more to prognosis at the time of detection.
That is, it is not clear from this type of data whether all
"early" disease will progress to "late" disease in the
absence of treatment (e.g., will all carcinoma in situ
progress to invasive cervical cancer) or how much "late"
disease goes through an "early" phase of any significant
duration (e.g., how much invasive cervical cancer arises
after a carcinoma in situ phase of some duration). Thus, to
support screening it is necessary to supplement these data
with some evidence, preferably a randomized controlled
trial, that earlier detection does improve mortality.
Again, because the diseases for which screening has been
undertaken are high prevalence diseases, it is possible to
collect such data. Finally, because of high incidence, age-
specific incidence rates can be determined with some
reliability. Thus, these high prevalence cancers provide

the data requirements necessary to develop the types of models discussed in the last section.

Also, because of high prevalence, risk factors for these cancers have been established by numerous epidemiological studies. For example, the association between breast cancer and family history (e.g., Haagenson, 1972) and age at first parity (MacMahon, et al. 1970) have been reported in the literature. In the case of colorectal cancer, diet (Burkett, 1971), geography (Wydner and Shigematsy, 1966), and precursor diseases (adenomatas polyps, ulcerative colitis, nitrosoamines) (Burkett, 1971) are all associated factors. For cervical cancer, epidemiological associations that have been proposed include ethnicity, early sexual activity with multiple partners and other risk factors (Kessler, 1974).

Risk factors identified by epidemiological studies may be useful in the development of cost-effective general population screening strategies. They provide a basis for stratifying population by risk categories and developing different strategies as a function of risk category, such that, given limited resources, some benefit measure can be optimized over the whole population. However, risk complexes of the types noted above are usually quite resistant to direct intervention with the intent of primary prevention of the disease. This is the case because such associations have as yet to be proven as necessary and/or sufficient causes of the disease in question (such as diet and colorectal cancer). Additionally, these risk complexes include significant behavioral components (age at first parity, etc.) with complex personal and social determinates that do not respond to public health interventions.

In conclusion, for the cancers for which general population screening seems feasible there is a substantial amount of clinical data and some data on risk factors, but relatively little data on etiology that offer promise for successful primary prevention. Thus, there are data to support development of disease models and analysis of population screening programs, but relatively little to support primary prevention activities.

B.  Industrial Cancers

The problems and policy issues related to occupational cancer are of a different sort. To differentiate occupational cancer from general population cancers, it is instructive to consider the general issues and the types of data that are available for informed decisions about screening industrial populations.

The issue of occupationally related cancer has become an increasingly important problem for any public strategy for cancer control or prevention. In the United States, among persons 15-65, 75% of men and 50% of women are gainfully employed. The occupational environment, then, becomes quite a significant focus for cancer studies.

During the last 20 years, examples of occupational
carcinogenesis have been increasingly recognized because of
extremely high risk ratios (often a function of the rarity
of the disease in the general population) and a very high
incidence rate among occupationally exposed persons, e.g.,
lung cancer in asbestos workers.

The particular question of the contribution of
occupational exposures to the overall incidence of cancer in
the United States is a important, but difficult question.
Estimates have varied from as low as 1% -5% (Wydner and
Gori, 1977) to as high as 20%- 40% (Bridbord et al., 1978).

If we raise questions about occupational cancer
screening analogous to the mass screening models described
above, we can discuss the generalizability of this approach
to consider occupationally based screening policies.
Generally, a particular strategy or policy for cancer
screening for any population should be dependent on both the
type and quality of available data concerning the disease
characteristics (natural history), population
characteristics (number at risk, age, etc.) and
characteristics of the screening technology (i.e.,
sensitivity and specificity).

Epidemiological studies in industrial populations have
documented very high relative risks for specific cancers in
particular occupational groups; lung cancer in shipyard
workers (Selikoff and Hammond, 1978) and leukemia in
petrochemical workers (Infante et al., 1977) are just two
examples from a large and growing body of literature.

The detection of extremely high risks for cancer in small localized industrial groups has followed from an initial clinical recognition of a rare cancer in a cluster of patients. Because these clusters are highly unusual events, specific factors are analyzed to explain the event. Often the most important of these factors is found to be occupational history, as in the paradigmatic example of mesothelioma associated with asbestos exposure in shipyard workers (Selikoff, 1976). Thus, from the recognition of a rare disease event, epidemiologic studies have investigated the type of occupational exposure in their search for a specific agent. These studies have discovered a large number of chemicals or industrial processes associated with excess cancer risk (Bridbord et al., 1978). There is, then, a large amount of data concerning etiologic factors in occupational cancer (carcinogens) that is more firmly established than the risk factor data available for most general population-based cancers. Thus, the emphasis of the current data base and the current research thrust is the examination of high-risk industrial groups by epidemiological studies and surveillance (Milham, 1976, Muir, 1976, Levy and Wegman, 1978).

On the other hand, there are far less data available that could be used to develop natural history models of occupational cancers. Because the problem of occupationally related cancer is recent, there are not sufficient historical data from clinical treatment records to consider rates of disease progression or survival probabilities.

This is even more the case for the rare, low-incidence
occupational cancers such as angiosarcoma of the liver or
mesothelioma, where the number of individuals with these
particular diseases has been small to date. (However, the
11 million persons who have had occupational exposure to
asbestos will, unfortunately, provide an extensive data base
for future development of an occupational disease natural
history.) As such, there is limited evidence that can be
used to develop models of occupational cancer analogous to
those developed for general population cancers. Rather, the
data that are available is on the carcinogenic potential of
certain substances. It is this information that offers the
potential for the development of different types of models,
specifically, models to screen substances.

C.  Potential of Mathematical Models to Plan Programs to
    Screen Workers

Because of the possibilities of primary prevention, earlier
detection of cancer in industrial groups is an important
goal. Thus, questions of appropriate screening strategies
for these groups should be considered.

The types of disease models discussed in Part II can be
used to consider the relative benefits of alternative
screening strategies for industrial workers. To do this, we
must make the assumption that the model of disease
progression, prognosis and screen reliability are similar in
the general population (usually the population for which

data are available for model development) and the industrial
population.  Analysis in an industrial population could then
be undertaken by considering industrial exposure as a risk
factor that increases age-specific incidence (i.e., $i_r(a)$ in
our simple model).

However, it may not be true that a model of disease
progression developed from general population data is
representative of some substance-induced disease.  General
population data are derived from a large heterogeneous group
exposed to a wide variety of potential carcinogenic agents.
Occupational induced cancers, on the other hand, usually
occur as the result of exposure to a single substance and
occur in a relatively homogenous occupational class.

Regardless of the applicability of any given disease
model, decisions about screening industrial populations must
be based on the types of underlying considerations
incorporated in these disease models.  That is, there is a
disease that progresses such that as the duration of
preclinical disease increases, prognosis declines; there is
a screening technique that can detect the disease earlier in
its course; and earlier detection improves mortality.
Without some evidence of the validity of these premises, the
high risk of industrial populations in and of itself
provides little basis for instituting screening programs.

So far, current attempts to develop screening programs
for highly at risk occupational groups have involved sputum
cytology for uranium workers exposed to lung cancer risks

and urine cytology for dyestuff workers exposed to lung
cancer risks. There have been no proven benefits in terms
of increased survival in any of the occupational cancer
screening programs to date, although disease can be detected
at an early stage (Greenburg et al., 1976, McEwan, 1976).
Those cancers for which there is more evidence of the value
of screening (i.e., breast, cervical and colorectal) are not
cancers that have been linked to any particular industrial
groups.

The importance of chemical carcinogenesis in industrial
populations and the large body of literature devoted to the
identification and definition of chemical carcinogenesis
raises the possibility of a different approach to cancer
screening programs - screening substances rather than
people. The methodological difficulties in developing this
type of screening program are numerous and current
techniques to screen substances have only been recently
developed. However, the potential to develop a program to
identify cancer-causing substances and limit human exposure
in established high risk groups makes this approach worthy
of further investigation. In what follows, we will briefly
describe some of the possible approaches to screening
carcinogens.

D.  Identification of Carcinogenic Substances in the
    Workplace

1.  *Epidemiological Studies and Animal Tests*

As indicated above, traditionally identification of
human carcinogens has been by epidemiologic methods.

However, this is a less than optimal approach to the problem
of risk identification for the following types of reasons:

1) Epidemiologic surveys in occupational settings are
   difficult and costly to conduct.  In most cases where
   specific substances and the related industrial process
   are associated with an excess of cancer incidence, the
   relative risks have been quite high.  Epidemiologic
   methods are not technically suited to data where the
   baseline frequencies of disease are small.

2) There is a large and diverse group of chemical used in
   industrial processes, 60,000 in common use and nearly
   2,000 new chemicals introduced each year.
   Epidemiological and case study approaches are not able to
   keep up with this rapid introduction of possible
   hazardous processes.

3) Occupational groups at risk for excess cancer are small,
   focused population clusters.  When a descriptive category
   is sufficiently small to make hypotheses about causal
   factors (like specific occupation and a particular
   cancer), other variables can become equally credible in
   an epidemiological analysis.  Well-defined cohorts of
   workers can also be expected to have a number of other
   traits in common such as diet, residential area, or
   cultural heritage.  It is therefore difficult,
   particularly in the case of small relative risks, to
   distinguish the occurrence of disease from background
   variation.  Also, because workers within an occupational

cohort change jobs frequently, it is extremely difficult
to make dose-response estimates on suspected carcinogens.

4) Most occupational carcinogensis is characterized by long
latent periods of up to twenty years between the time of
exposure to a substance and the clinical surfacing of a
tumor.  Hence, if reliance is placed on human
epidemiological evidence, there will have been a great
deal of exposure to the chemicals before carcinogenic
potential is determined.

Traditionally, the carcinogenicity of a suspected
chemical agent not identified by human epidemiologic
analysis has been established when administration of the
agent to animals in an adequately designed and conducted
experiment resulted in an increased incidence of one or more
types of cancer in the test animals.  Such increases are
regarded with greater confidence if positive results are
observed in more than one group and species of animals.
Also, the evidence of a dose-response relationship
strengthens the evidence for carcinogenesis (Tomatis, 1976).
However, extrapolation from animal studies to estimate human
risks entails uncertainty, due mainly to questions
concerning duration of exposure, route of administration,
metabolism (including species variation), host
susceptibility and relation to usual human exposure, and low
dose extrapolation (see Chapter 9).

Though for some time animal tests will no doubt provide
the principal criteria by which carcinogenicity is

determined, the "utility of animal cancer tests for cancer prevention is limited by several important factors" (Ames, 1979): they are extremely expensive (about $200,000 per test), take a long time to complete (about 3 years), require pathologists (who are in limited supply), and are not good at identifying carcinogens in complex chemical mixtures. Hence, it is important to develop some means of at least prioritizing chemicals for bioassay and perhaps to guide policy while awaiting the results of bioassay.

An approach to substance screening would be most easily obtained if certain a priori properties of a chemical "marked" it as a carcinogen. The search for this kind of evidence concerning the structure-activity relationships in carcinogenic substances has not established a direct approach to chemical classification without the aid of some biological test systems (Newman and Huns, 1977). Therefore because of the only rudimentary understanding of the relationship between clinical structure and the possible mutagenicity or carcinogenicity of a substance, models of structure-activity-based screening approaches are not possible at this time.

## 2. *In Vitro Tests*

During the last decade, a new approach to screening chemical substances has been developed partially in response to the problems engendered by traditional bioassays. Short-term tests, based on the relationships between DNA damage in the cell and subsequent carcinogenesis, have been developed

in several *in vitro* systems. These tests hold promise as a
potentially useful first line screen for suspected chemical
substances, and many are already used for this purpose. The
most widely studied of the tests presently in use is the
Salmonella test for mutagenic assay devised by Ames (1979).
This test is representative of the *in vitro* approach and
will be discussed briefly.

The test involves combining Salmonella bacteria of a
particular testing strain on a petri dish with the suspected
chemical compound. Homogenates of liver are also placed in
the dish to provide the enzymes necessary for metabolic
activation of the potential carcinogen. Test results are
usually obtained after two days incubation by counting the
number of mutated strains and generating a dose-response
curve. The greater the number of bacterial mutations found
in the test, the more potent a carcinogen is considered to
be (Ames, 1979).

McCann *et al..*(1975) have conducted for the Salmonella
test the type of validity studies that are necessary for all
short-term tests. In their examination of 300 chemicals
that had been established as carcinogens or non-carcinogens
in animal tests, they determined that 90% of the chemical
carcinogens were, in fact, mutagenic in the Salmonella test.
About 10% of non-carcinogenic substances appeared as
mutagens in the test. The Ames test has also been
independently evaluated by Purchase *et al.* (1976). This
study analyzed six different short-term tests in terms of

their sensitivity to carcinogenicity in 120 selected
compounds.  The Ames test had a false positive rate of 7%
and a false negative rate of 9% in this trial.  These
results have been further confirmed in independent test in
other laboratories (de Serres, 1979).  However, Rinkus and
Fegator (1980) have demonstrated that the reliability of the
Ames test varies for different chemical classes.

There are other short-term systems that have been
evaluated for their potential as carcinogenesis screens.
Hollstein *et al.* (1979) and others have reviewed many of the
major short-term tests.  Because certain of these tests
respond better to particular classes of chemical substances,
a combination of validity studies for each particular test
and "matrix" studies to establish the relationship between
the various tests is an important first step in developing a
comprehensive screening approach for chemicals.

There are several potential advantages in the short-
term testing approach to substance screening.  These tests
provide versatile tools for the screening of the complex
chemical mixes common to industrial exposures.  The cost of
*in vitro* testing is low relative to animal testing,
averaging between $200 and $2,000 per test depending on the
complexity of the particular cell system chosen, the time to
perform the test is relatively short, and in some cases test
sensitivity is relatively high.  However, since none of the
tests can establish with certainty whether a compound will
be carcinogenic in either experimental animals or humans,

and since different tests have different reliability for
different classes of compounds, there are problems involved
in making inferences about carcinogenicity from *in vitro*
test results.  Further, the reliability of the tests, the
validity across practitioners and the power of the tests
have not been well established.  However, because each of
the testing systems responds differently to particular
classes of chemicals, it may be possible to validate the
strength and weakness of the various tests with the aim of
the development of a battery of short-term tests chosen to
provide the greatest amount of information in the fewest
tests.  The outcome of these short-term test series could at
least provide information concerning the priority that
should be assigned to further *in vivo* tests or, more
optimistically, reduce our current dependence on the lengthy
and expensive animal tests.

There have been proposals suggesting series of short-
term tests to detect carcinogens (Williams, 1980, summarizes
several of these proposals).  These strategies have been
based on the use of a range of tests that attempt to
identify substances operating through the different
mechanisms of carcinogenicity.  They consider the predictive
capacity of each test, the theoretical basis for each test
(which relates the end point of the test to carcinogenicity)
and the reproducibility of the test.  The proposals have
been for a battery of tests or the use of a tier approach,
where results at one tier determine whether a substance will

be evaluated at the next tier. However, none of these proposals have considered explicitly the cost of the strategy versus the gain in information content, compared to some alternative strategy.

## E. Models for Analyzing Screening Strategies to Identify Potential Carcinogens

### 1. *Overview*

When consideration is given to screening people, the relative benefits of alternative screening strategies are dependent to a large extent on the rate at which prognosis at treatment declines as a function of increases in the duration of preclinical disease. This is the reason modelling efforts have focused on underlying disease dynamics, i.e., deep models have been developed. Also, for the different cancers, there are a very limited number of screening techniques that can be used and often, when more than one technique is available, the order in which to give the techniques at an examination is not an issue (due to relative costs, for example). Hence, the design of what is actually performed at a screening examination has received little attention (for an exception, see Neuhauser and Lewicki (1975) and Cornell (1978)). Finally, because of the extent to which treatment for the disease has already been analyzed and efficacy determined, it is possible to focus on screening strategies as a problem relatively separate from an analysis of treatment strategies.

When one considers the screening of substances, a
different situation exists.  The underlying condition for
which the search is made i.e., the carcinogenic potency of
the substance, does not change in time.  However, there are
many tests that can be given, each with different
reliabilities for different classes of chemicals, and each
with different costs.  This suggest that the important
question in substance screening has to do with which test to
include in a screen, in what order to give the tests, and
how to interpret results.

One result of an analysis of a substance screening
strategy would be some indication of the carcinogenic
potency of a substance as a function of tests performed.  It
is less clear how to integrate this result into the overall
decision process than the comparable type of result from
analysis of population screening.  To make decisions about
testing strategies for carcinogenicity we have to have some
way of evaluating the benefit of changes in certainty about
carcinogenic potency.

Further, since there is no experience comparable to the
data on the results of treatment of disease, it is difficult
to consider screening of substances independently of the
broader issue of toxic substance regulation.  Thus, we are
led to questions such as how many people are exposed to the
substance, what are the costs and effectiveness of limiting
worker exposure to the substance, and who is going to bear
these costs.  Because these costs are often quite high and

because they fall on powerful groups, the debate about toxic
substance regulation has been quite acrimonious.

In this environment, what is needed is a framework that
will help structure this debate, as well as provide some
basis for the types of modelling activities necessary to aid
decision-making.  It is our feeling that decision analysis
(Raiffa, 1968, DeGoot, 1970) offers such a framework.  One
advantage of a decision analytic framework is that it
encourages simultaneous consideration of the characteristics
of tests used to screen substances and both the potential
costs and benefits (derived from extent of worker exposure
and estimated carcinogenic potential of the substance) of
reducing exposure to the substance.  This framework suggests
what Weinstein (1979) has called a parallel system of
information development, where data on toxicity, exposure
and control costs are developed simultaneously but
progressively more accurately, as opposed to a series
system, where data are developed first on toxicity, then
exposure, then control costs.  For only in a parallel system
of information development is it possible to determine the
cost of either a false-positive or false-negative test
result and hence the value of changing the design of the
test sequence to change the likelihood of these results.

A further advantage of a decision analytic framework is
that an explicit distinction can be made between questions
of probabilities and questions of values.  Questions about
probabilities are of the following type:  how likely is a

chemical to be a carcinogen of some potency, given available
information about the chemical. Value questions are of the
following type: given the number of people exposed to the
chemical, the probability the chemical is a carcinogen of
some potency, and the costs of alternative regulatory
action, what level of regulation is appropriate? Whereas
questions about probabilities involve interpretation of
data, value questions involve tradeoffs between dollar costs
and expected cases of cancer. (This distinction is implied
in a recent report by the Office of Sciences and Technology-
Identification, Characterization and Control of Potential
Human Carcinogens: A Framework for Federal Decision-Making
(1979). In this report the distinction is phased in terms
of questions of science versus questions of regulation.)

As mentioned, proposals for testing strategies to
identify carcinogens have not been based upon an explicit
consideration of the costs of the tests versus the value of
information gained from the test. One reason for this has
been the lack of information on the relationship between the
false-positives and false-negatives on different tests,
information which is expensive to gather. However, there
are extensive efforts worldwide to evaluate individual and
combinations of *in vitro* tests and hence, there may soon be
a sufficient data to begin to analyze the value of
alternative test sequences (e.g., Kawachi, 1980).

Weinstein (1979) and Page (1981) use a decision
analytic framework as the basis for discussing and

illustrating issues in the design of cost-effective
screening strategies for carcinogenicity. In what follows,
we provide a further illustration of the approach.

2.  *Analysis of Substance Testing Strategies With In Vitro Tests*

    Assume that for chemicals of a certain type, a
preliminary assessment of extent of worker exposure and
costs of limiting exposure is made. Based on these results,
a decision like the following is made: If there is evidence
that there is at least a 70% chance a chemical will be
positive on animal bioassay, worker exposure to the chemical
will be limited until bioassay results are available; if
evidence indicates under 0.5% chance of carcinogenicity the
chemical will not be considered for animal bioassay. (To
simplify this example we will ignore the potential potency
of the substance.) It should be emphasized that the
relationship of 70% and 0.5% to the extent of worker
exposure and the costs of control are value decisions
resulting from subjective tradeoffs between the costs of
limiting exposure and the costs of cancer incidence. There
is little scientific evidence to help in determining whether
these percentages should be 60% or 80%, or 0.25% or 1%.
However, evidence can be marshalled to determine for any
given chemical whether the chance it will be positive in
animal bioassay is greater than 70% or less than 0.5%. In
what follows we demonstrate how such evidence would be

derived when there is only one short-term test that can be given and when there are two tests that can be given.

a. *Single Test Analysis.* Because the Ames test is the *in vitro* test about which most information is available, we will consider this as a prototype test. We will be interested in the following: (1) given a substance is positive on the Ames tests, what is the probability it will be positive (i.e., a carcinogen) on animal bioassay; (2) given a substance is negative on the Ames test, what is the probability it will be negative (i.e., not a carcinogen) on animal bioassay. Let

C = event substance is positive on animal bioassay,

C' = event substance is negative on animal bioassay,

A = event substance is positive on the Ames test,

A' = event substance is negative on the Ames test.

Then, we are interested in $P(C|A)$ and $P(C'|A')$. From experimental data (Ames, 1979) we know that for certain classes of chemicals (we will assume the chemical under consideration comes from one of these classes) $P(A|C) = .9$ and $P(A'|C') = .8$. Therefore, $P(A'|C) = .1$ and $P(A|C') = .2$. Using Bayes Theorem,

$$P(C|A) = \frac{P(A|C)P(C)}{P(A|C)P(C) + P(A|C')P(C')}$$

and

$$P(C'|A') = \frac{P(A'|C')P(C')}{P(A'|C')P(C') + P(A'|C)P(C)}.$$

To determine $P(C|A)$ and $P(C'|A')$, we need $P(C)$, that is the prevalence of chemicals considered for animal bioassay that would be positive on animal bioassay. Historically, about 20% of chemicals tested on animals in adequately designed experiments have been positive. Since we have no reason to suspect our current ability to identify potential carcinogens is any worse than it has been historically, we might estimate a lower bound for P(C) as .20. For any given chemical, we might estimate, perhaps based on structure-activity relationships, a higher probability of positive animal results. However, if what follows, we will assume P(C) = .20. Thus

$$P(C|A) = \frac{.9P(C)}{.9P(C) + .2P(C')} = \frac{.9P(C)}{.7P(C) + .2}.$$

If P(C) = .2, P(C|A) = .53. Thus, at a minimum, a chemical that is positive on the Ames test has slightly over a 50% chance of being an animal carcinogen. Since this value is less than 70%, if the Ames test is the only adequate validated test, a positive result does not provide strong enough indication of carcinogenicity to limit worker exposure while awaiting the results of animal bioassay.

We can go through the same type of reasoning to determine that P(C'|A') = .97 for P(C) = .20. That is, when P(C) = .20 there is a 3% chance that the substance is a carcinogen even though it is negative on the Ames test. Since this is greater than .5%, if the Ames test is the only test available, a negative result does not provide sufficient evidence to eliminate the chemical from consideration for bioassay.

*b. Analysis for Two Tests.* Assume we want to consider the
role of a second test as part of a testing sequence to be
used to identify carcinogens, where the Ames test is the
first test. Let X = event that the second test is positive
and X' = event that the second test is negative. We are
interested in P(C|AX). Specifically, for some type of
policy decision we have determined P(C|AX) must be greater
than Z. The problem we will address here is the following:
given the characteristics of the Ames test, how do we
evaluate the second test and how do we determine whether the
second test will provide sufficient additional information
to make our decision.

From Bayes' formula,

$$P(C|AX) = \frac{P(AX|C)P(C)}{P(AX|C)P(C) + P(AX|C')P(C')}.$$

Also, P(AX|C)P(C) = P(X|AC)P(A|C)P(C). Thus,

$$P(C|AX) = \frac{P(X|AC)(.9)P(C)}{P(X|AC)(.9)P(C) + P(X|AC')(.2)P(C')}.$$

By manipulating the algebra and substituting Z for P(C|AX),
it can be shown that

$$\frac{P(X|AC)}{P(X|AC')} = \frac{.2P(C')Z}{.9P(C)(1 - Z)}.$$

This equation tells us that if we want to evaluate the
role of a second test to be used in conjunction with the
Ames test, we should evaluate the test on two classes of
chemicals: one, known carcinogens (i.e., chemicals positive
on bioassay) that are also positive on the Ames test; two,

known non-carcinogens (i.e., chemicals negative on bioassay)
that are positive on the Ames test. By focusing on the
ratio of the percentage of chemicals positive on the second
test in each of these two classes, we can determine for any
level of Z and any a priori insight into the prevalence of
carcinogenic substances in a particular environment, whether
the second test provides sufficient information to warrant
its use if a substance is positive on an Ames test.

For our example, assume we have evaluated this second
test and determined the ratio $P(X|AC)/P(X|AC')$. The
question becomes, is this ratio sufficiently high to
recommend a second test? That is, if the chemical is
positive on the second test will $P(C|AX) > .7$? If not, then
there would be no value in giving the second test. If we
assume $P(C) = .2$, then if the characteristics of the second
test are such that the ratio is greater than 2.1, we should
give the second test. If the chemical was positive on the
second test, we would limit worker exposure while awaiting
the results of animal bioassay.

Again, going through similar reasoning, we can determine
the relevant ratio for deciding when there is some
probability, Z, a product is not carcinogenic.
Specifically,

$$\frac{P(X'|A'C')}{P(X'|A'C)} = \frac{.1Z\ P(C)}{.8(1 - Z)P(C')} .$$

In our example, for Z = .995 and P(C) = .20, this ratio
should be 6.2.

c. *Summary.* In this example, we have attempted to
demonstrate how decision analysis can provide a framework
for considering issues in toxic substance regulation. It is
important to note the distinction between probabilities and
values in this approach. Determination of the probability
of carcinogenicity can be made relatively independently of
value considerations. Likewise, value decisions can be made
independently of experimental results. However, decision-
making involves integration of these two considerations.
For, without value decisions we have no basis for
interpreting experimental results. And, having only
determined the probability of carcinogenicity, we have in no
sense resolved the decision-making or regulatory issues.


F.  Conclusion

    In this section, we have attempted to demonstrate the
following:

1)  Though models developed to analyze screening strategies
    in the general population can be used to analyze
    screening strategies in industrial populations, the
    cancers of high prevalence in the general population,
    those for which models have been developed, are not of
    any greater importance in specific industrial
    populations. The important industrial related cancers
    have been of relatively low incidence in the general
    population, and hence there are little data available

for model development, or they are cancers that so far
are not amendable to reduction in mortality through
screening.

2) The "cancer problem" in industrial populations lends
itself to primary prevention activities more than does
the "cancer problem" in the general population. This
suggests a focus on identifying the carcinogenic potency
of a substance and limiting worker exposure, that is, a
focus on toxic substance regulation.

3) The nature of the decision problems involved in toxic
substance regulation are more complex than the decision
problems involved in screening people. This is due
primarily to the high potential costs of regulatory
action and to concerns about who bears what kinds of
costs. These issues are often contested politically.
As a result, it is important to have a conceptual
framework to consider various aspects of this problem.

4) We consider decision analysis to be a useful framework
for posing and analyzing a variety of issues in toxic
substance regulation. It is useful because it
integrates test characteristics and the cost of
different types of errors and it makes more explicit the
distinction between questions of probabilities (i.e.,
scientific questions) and questions of value (i.e.,
regulatory questions) and the necessity of their
integration in decision-making. Further, it provides a
basis for analyzing probabilistic questions.

SHWARTZ AND PLOUGH

Albert, A., Gertman, P., and Louis, T. (1978 a). Screening
for the early detection of cancer - I. The temporal
natural history of a progressive disease state.
*Mathematical Biosciences, 40*: 1-59.

Albert, A., Gertman, P., Louis, T., and Liu, S. (1978 b).
Screening for the early detection of cancer - II.
The impact of screening on the natural history of
the disease. *Mathematical Biosciences, 40*: 61-109.

American Cancer Society. (1980). Report on the cancer
related health checkup. *CA, 30*: 194-240.

Ames. B. (1979). Identifying environmental chemicals
causing mutations and cancer. *Science, 204*:
587-593.

Blumenson, L.E. (1976). When is screening effective in
reducing the death rate? *Mathematical Biosciences,
30*: 273-303.

Blumenson, L.E. (1977 a). Detection of disease with
periodic screening: Transient analysis and
application to mammography examination.
*Mathematical Biosciences, 33*: 73-106.

Blumenson, L.E. (1977 b). Compromise screening strategies
for chronic diseases. *Mathematical Biosciences, 34*:
79-94.

Blumenson, L.E. and Bross, I.D.J. (1969). A mathematical
analysis of the growth and spread of breast cancer.
*Biometrics, 22*: 95-109.

Bridbord, K., Decoutle, P., Fraumeni, J.F., Hoel, D.G.,
Hoover, R.N., Rall, D.P., Saffiotti, U.,
Schneiderman, M.A., and Upton, A.C. (1978).
*Estimates of the Fraction of Cancer in the United
States Related to Occupational Factors.* National
Cancer Institute, Bethesda.

Bross, I.D.J., Blumenson, L.E., Slack, N.H., and Priore,
R.L. (1968). A two disease model for breast cancer.
In *Prognostic Factors in Breast Cancer*
(A.P.M. Forrest and P.B. Bunkler, eds.). Williams
and Wilkins, Baltimore, pp. 288-300.

Burkitt, D.P. (1971). Epidemiology of cancer of the colon
and rectum. *Cancer, 78*: 3-13.

Coppleson, L.W. and Brown, B. (1975). Observations on a
    model of the biology of carcinoma of the cervix:   a
    poor fit between observation and theory. *American
    Journal of Obstetrics and Gynecology*, 122: 127-136.

Coppleson, L.W. and Brown, B. (1976). The prevention of
    carcinoma of the cervix. *American Journal of
    Obstetrics and Gynecology*, 125: 153-157.

Cornell, R.G. (1978). Sequence length for repeated
    screening tests. *Journal of Chronic Diseases*, 31:
    539-545.

Degroot, M.H. (1970). *Optimal Statistical Decisions*.
    McGraw-Hill, New York.

de Serres, F.J. (1979). Evaluation of tests for
    mutagenicity as indicators of environmental mutagens
    and carcinogens. *Annals of the New York Academy of
    Science*, 329: 75-84.

Drake, J.W., de Serres, F.J., Darby, W.J., Dunkee, V.C.,
    Longfellow, D.G., Off, H., Purchase, I.F.H., Ramel,
    C., Schlatter, C., and Smith, E.M.B. (1980). Report
    11: Rationale for development of short-term assays
    for evidence of carcinogenicity. In *Long-Term and
    Short-Term Screening Assays for Carcinogens:   A
    Critical Appraisal*. IARC Monograph on the
    Evaluation of the Carcinogenic Risk of Chemicals to
    Humans, IARC, Lyon.

Dubin, N. (1979). Benefits of screening for breast cancer:
    Application of a probabilistic model to a breast
    cancer detection project. *Journal of Chronic
    Diseases*, 32: 145-151.

Dubin, N. (1981). Predicting the benefit of screening for
    disease. *Journal of Chronic Diseases*, 18: 348-360.

Eddy, D.M. (1977). *Rationale for the Cancer Screening
    Benefit Program Screening Policies: Implementation
    Plan, Part III*. Report to the National Cancer
    Institute, Blue Cross Association, Chicago.

Eddy, D.M. (1980). *Screening for Cancer: Theory, Analysis
    and Design*, Prentice-Hall, Inc., Englewood Cliffs,
    New Jersey.

Eddy, D.M. and Shwartz, M. (1982). Mathematical models in
    screening. In *Cancer Epidemiology and Prevention*
    (D. Schottenfeld and J.F. Fraumeni, Jr., eds.)
    W.B. Saunders and Co., Philadelphia.

Ellwein, L.B. (1979 a). *Bladder cancer with interventions: A computerized simulation model, Part I: Model development and Part II: Experimentation.* Science Applications, La Jolla, California.

Ellwein, L.B., Friedell, G.H., Greenfield, R.E., and Hilgar, A.G. (1979 b). Disease process modeling as a tool for planning and strategy assessment. *Seminars in Oncology, 6:* 260-264.

Galliher, H.P. (1969). *Optimal Ages for Pap Smears - A Preliminary report.* Department of Industrial Engineering, University of Michigan, Ann Arbor, Michigan.

Galliher, H.P. (1976). *Cost-Effective Planned Lifetime Schedules of Pap Smears: Estimated Maximal Potentials.* The Michigan Cancer Foundation, Detroit.

Galliher, H.P. (1977). *Optimal Ages for Pap Smears Using a Multi-Strain Model of Cervical Cancer.* The Michigan Cancer Foundation, Detroit.

Galliher, H.P. (1981). Optimizing ages for cervical smear examinations in followed healthy individuals. *Genecologic Oncology, 12:* 5188-5205.

Gohagan, J.K. and Swift, J.G. (1981). Scheduling pap smears for asymptomatic women. *Preventive Medicine, 10:* 741-753.

Gohagan, J.K., Rodes, N.D., Ballinger, W.K., Blackwell, C.W., Butcher, H.R., Darby, W.P., Pearson, D.K., Spitznagel, E.L., and Wallack, M. (1982). *Early Detection of Breast Cancer: Risk, Detection Protocols, and Therapeutic Implications.* Praeger, New York.

Greenburg, D.S., Hurst, G.A., Matlage, W.T., Miller, J.M., Hurst, I.J., and Mabry, L.C. (1976). Tyler asbestos workers program. In *Occupational Carcinogenesis,* (U. Saffiotti and J.K. Wagoner, eds.). New York Academy of Science, New York, pp. 353-364.

Haagensen, C.D. (1972). Family history of breast carcinoma in women predisposed to develop breast carcinoma. *Journal of the National Cancer Institute, 48:* 1025-1027.

Habbema, J.D.F., van Oortmarssen, G.J., and Lubbe, J.Th.N. (1980). *Simulation Model for Evaluation of Mass Screening.* Department of Public Health and Social Medicine, Erasmus University, Rotterdam, The Netherlands.

Hollstein, M., McCann, J., Angelosanto, F., and Nichols,
     W. (1979). Short-term tests for carcinogens and
     mutagens. *Mutation Research*, 65: 133-226.

Infante, P.F., Rinsky, R.A., Wagoner, J.K., and Young,
     R.J. (1977). Leukemia in benzene workers. *Lancet*,
     July 9, 1977, 76-78.

Kawachi, T., Komatsu, T., Kada, T., Ishidate, M., Sasaki,
     M., Sugiyama, T., and Tazima, Y. (1980). Results of
     recent studies on the relevance of various short-
     term screening tests in Japan. In *Predictive Value
     of Short-Term Screening Tests in Carcinogenicity
     Evaluation* (G.M. Williams, R. Kroes, H.W. Waaijers,
     and K.M. van de Poll, eds.). Elsevier/North
     Holland, Amsterdam, pp. 253-268.

Kessler, I.I. (1974). Perspectives on the epidemiology of
     cervical cancer with special reference to the herpes
     virus hypothesis. *Cancer Research*, 34: 1091-1110.

Kirch, R. and Klein, M. (1978). Prospective evaluation of
     periodic breast examination programs. Interval
     cases. *Cancer*, 41: 728-736.

Knox, E.G. (1973). A simulation system for screening
     procedures. In *Future and Present Indicatives,
     Problems and Progress in Medical Care, Ninth Series*.
     (G. McLachlan, ed.). Nuffield Provincial Hospitals
     Trust, Oxford, London, pp. 17-55.

Knox, E.G. (1975). Simulation studies of breast cancer
     screening programmes. In *Probes for Health*. Oxford
     University Press, London, 13-44.

Levy, B.S. and Wegman, D.H. (1978). Physician-based
     surveillance of occupational lung disease.
     Presented at the Annual Meeting of the American
     Public Health Association, Los Angeles.

Lincoln, T. and Weiss, G.H. (1964). A statistical
     evaluation of recurrent medical examination.
     *Operations Research*, 12: 187-205.

Louis, T.A., Albert, A., and Heghinian, S. (1978). Screening
     for the early detection of cancer - III. Estimation
     of disease natural history. *Mathematical
     Biosciences*, 40: 111-144.

MacMahon, B., Cole, P., Lin, T.M., Lowe C.R., Mirra, A.P.,
     Ravnihar, B., Salber, E.J., Valaoras, V.G., and
     Yuasa, S. (1970). Age at first birth and breast
     cancer risk. *Bulletin of the World Health
     Organization*, 43: 209-221.

McCann, J. and Ames, B.N. (1976). Detection of carcinogens
    as mutagens in the salmonella microsome test: Assay
    of 300 chemicals. Discussion. *Proceedings of the
    National Academy of Science, USA*, 73: 950-960.

McCann, J., Choi, E., Yamasaki, E., and Ames, B.N. (1975).
    Detection of carcinogens as mutagens in the
    salmonella microsome test: Assay of 300 Chemicals.
    *Proceedings of the National Academy of Science, USA*,
    72: 5135-5139.

McEwan, J.C. (1976). Cytological monitoring of nickel
    sinter plant workers. In *Occupational
    Carcinogenesis*, (U. Saffiotti and J.K. Wagoner,
    eds.). New York Academy of Science, New York,
    pp. 365-369.

Milham, S. (1976). *Occupational Mortality in Washington
    State, 1950-1971, Volumes I-III*. U.S. Department of
    Health, Education and Welfare, H.E.W. Publication
    No. (NIOSH) 76-175.

Muir, C.S. (1976). Feasibility of monitoring populations to
    detect environmental carcinogens. *Inserm Symposia
    Series*, 52: 279-294.

Neuhauser, D. and Lewicki, A.M. (1975). What do we gain
    from the sixth stool guaiac? *New England Journal of
    Medicine*, 293: 226-228.

Newman, M.S. and Hung, W.M. (1977). Structure-carcinogenic
    activity relationships in the Benz (a) anthracene
    Series. 1, 7, 12- and 2, 7, 12- Trimethylbenz (a)
    anthracenes. *Journal of Medicinal Chemistry*, 20:
    179-181.

Page, T. (1981). A framework for unreasonable risk in the
    toxic substances control act (TSCA). *Annals of the
    New York Academy of Sciences*, 363: 145-166.

Prorok, P.C. (1976 a). The theory of periodic screening I:
    Lead time and proportion detected. *Advances in
    Applied Probability*, 8: 127-143.

Prorok, P.C. (1976 b). The theory of periodic screening II:
    Doubly bounded recurrence times and mean lead time
    and detection probability estimation. *Advances
    Applied Probability*, 8: 460-476.

Purchase, I.F.C., Longstaff, E., Ashby, J., Styles, J.A.,
    Anderson, D., Lefevre, P.A., and Westwood,
    F.R. (1977). Evaluation of six short-term tests for
    detecting organic chemical carcinogens and
    recommendations for their use. *Nature*, 264:
    624-627.

Raiffa, H. (1968). *Decision Analysis: Introductory Lectures on Choices Under Uncertainty*. Addison-Wesley, Reading, Massachusetts.

Rinkus, S. and Legator, M.S. (1980). The need for both *in vitro* and *in vivo* systems in mutagenicity screening. In *Chemical Mutagens: Principles and Methods for Their Detection* (F.J. de Serres and A. Hollaender, eds.). Plenum Press, New York, pp. 365-372.

Selikoff, I.J. (1976). Lung Cancer and Mesothelioma During Prospective Surveillance of 1249 Asbestos Insulation Workers, 1963-1974. In *Occupational Carcinogenesis*, (U. Saffiotti and J.K. Wagoner, eds.). New York Academy of Science, pp. 448-457.

Shwartz, M. (1978 a). An analysis of the benefits of serial screening for breast cancer based upon a mathematical model of the disease. *Cancer, 41*: 1550-1564.

Shwartz, M. (1978 b). A mathematical model used to analyze breast cancer screening strategies. *Operations Research, 26*: 937-955.

Shwartz, M. (1980). Estimates of lead time and length bias in a breast cancer screening program. *Cancer, 46*: 844-851.

Shwartz, M. (1982). Validation and use of a mathematical model to estimate the benefits of screening younger women for breast cancer. *Cancer Detection and Prevention, 4*: 595-601.

Swanson, M.G. and Shaw, E. (1978). A behavioral model for community cancer control. Presented at the American Association for the Advancement of Science Annual Meeting, Washington, D.C.

Thompson, D.E. and Disney, R. (1977). *A Mathematical Model of Progressive Diseases and Screening*. The Michigan Cancer Foundation, Detroit.

Thompson, D.E. and Doyle, T.C. (1976). *Analysis of Screening Strategies for Colo-rectal Cancer: A Description of an Approach*. The Michigan Cancer Foundation, Detroit.

Thompson, D.E. and Jacobi, L.W. (1977). *Analysis of Screening Strategies for Colo-rectal Cancers: An Extension of the Model*. The Michigan Cancer Foundation, Detroit.

Tomatis, L. (1976). The I.A.R.C. program on the evaluation
    of carcinogenic risk of chemicals to man. *Annals of
    the New York Academy of Science*, 271: 396-409.

van Oortmarssen, G.J. (1979). *Models in Mass Screening for
    Cervical Cancer - A Progress Report*. Department of
    Public Health and Social Medicine, Erasmus
    University, Rotterdam, The Netherlands.

van Oortmarssen, G.J., Habbema, J.D.F., Lubbe, J.Th.N., de
    Jong, G.A., and van der Maas, P.J. (1980).
    Predicting the effects of mass screening for disease
    - a simulation approach. *European Journal of
    Operations Research*, 5: 461-471.

Voelker, J.A. (1976). *Contributions to the Theory of Mass
    Screening*: Department of Industrial Engineering and
    Management Sciences. Northwestern University,
    Evanston, Illinois.

Weinstein, M.C. (1979). Decision making for toxic
    substances control: Cost-effective information
    development for the control of environmental
    carcinogens. *Public Policy*, 27: 333-383.

Weisburger, J.H. and Williams, G.M. (1981). Carcinogen
    testing: Current problems and new approaches.
    *Science*, 214: 401-407.

Williams, G.M. (1980). Batteries of short-term tests for
    carcinogen screening. In *Predictive Value of Short-
    Term Screening Tests in Carcinogenicity Evaluation*
    (G.M. Willliams, R. Kroes, H.W. Waaijers, and K.M.
    van de Poll, eds.). Elsevier/North Holland,
    Amsterdam, pp. 327-346.

Wynder, E.L. and Gori, G.B. (1977). Guest editorial:
    Contributions of the environment to cancer
    incidence: an epidemiological exercise. *Journal of
    the National Cancer Institute*, 58: 825-832.

Wynder, E.L. and Shigematsy, T. (1967). Environmental
    factors of cancer of the colon and rectum. *Cancer*,
    20: 1520-1561.

Zelen, M. and Feinleib, M. (1969). On the theory of
    screening for chronic disease. *Biometrics*, 56:
    601-614.

# 9
# THE ANALYSIS OF ANIMAL
# CARCINOGENICITY EXPERIMENTS

Richard G. Cornell
Robert A. Wolfe
William J. Butler

The University of Michigan
Ann Arbor, Michigan

## I. INTRODUCTION

The need for screening of substances for carcinogenicity was
discussed in Chapter 8. In this chapter we will review
statistical methods and issues related to animal experiments
which are designed to screen substances for carcinogenic
potential. The need to evaluate new substances before large
numbers of people are exposed to them is a major
justification for the use of animal experiments for such
screening purposes. Animal studies are also useful for
studying the carcinogenic effects of substances that are
already in the environment. Epidemiologic studies of such
substances are sometimes impractical because they would be

prohibitively expensive or complex. In addition, in human studies, there is generally no experimental control over exposure levels and the precise measurement of exposure levels is often difficult. Animal studies, in contrast, are relatively inexpensive and allow for precise control of levels of exposure to the tested substance.

The statistical analysis of animal experiments for carcinogenicity is addressed in depth in this chapter, although such analysis is just one component of the screening process. Toxicologists, pathologists, animal scientists, pharmacologists and statisticians often work as a team to interpret the results of such experiments. The results of such experiments are of use both in furthering our scientific understanding of carcinogenesis and in forming the basis for regulating the use of substances in the human environment.

A major goal of the regulatory process for substances is to limit the potential for the development of cancer in humans while retaining some of the benefits to society from their use. In many cases, the regulatory process is directed towards deciding whether to allow or prevent the introduction of substances into the environment. This emphasis on decision making makes the hypothesis testing component of statistical analysis relatively important, although the estimation of risk is also important for the regulation of levels of exposure.

Lifetime animal studies have been used extensively for screening compounds for both carcinogenicity and other forms of toxicity. Typically, these experiments involve exposure of small rodents to several doses of the substance being tested with determination of the presence or absence of tumors being made at the time of the animal's death or sacrifice. Analysis of the relationship between dose and tumor incidence or prevalence is the major focus of the statistical analysis of the experiments. Results of such analyses are often of central importance in reaching decisions about the control of human exposure to the tested substance.

Methods of analysis based on lifetime tumor rates are discussed in Section II. One of the main statistical issues which arises in such analyses is the choice of an appropriate analytical unit. Since many potential sites for tumors are examined for each animal, rates could be studied for each site separately, or a statistic summarizing information on all sites could be computed for each animal and incorporated into the analysis of the overall experiment. Once a choice of the basic unit and statistic is made, then analytic procedures which make maximal use of such summary statistics are needed. In particular, analysis is usually carried out at several sites, and an important issue is whether or not these separate analyses should be combined into an overall analysis or whether or not the interpretation of separate analysis should take into account the other sites studied in a formal manner.

In addition, when experimental animals are grouped by litter, the issue arises of whether or not this grouping should be taken into account. If substantial variability in incidence rates exists among litters, then the litter is the appropriate unit of analysis. A method of analysis which takes the intra-litter correlations into account should then be used.

Another issue which arises in the analysis of tumor rates is the appropriate method for describing the pattern of rates over different dosages. This is particularly important because these results might be used to estimate the carcinogenic risk to man through extrapolation from the high doses used in animal studies to lower doses which humans are more likely to encounter. Such extrapolations are the subject of Section IV.

Statistical methods for the analysis of time to tumor development are discussed in Section III. They address the fact that experimental data may be relevant to estimation of tumor prevalence or to estimation of tumor incidence. Statistical methods should therefore use data on tumors discovered either at the time of sacrifice or at the time of natural death, as well as data on animals exhibiting no evidence of tumors. Methods for the analysis of censored data are of prime importance. An issue which arises is the appropriate method of analysis in two extreme situations:

one in which the type of tumor under study is viewed as the
cause of death and the second in which the tumor is viewed
as incidental to the death. Related issues arise in
determining appropriate methods of analysis for experimental
designs which involve the sacrifice of subgroups of animals
at successive time points during the experiment.

Approaches for the resolution of these statistical
issues are presented in this chapter. Many articles on
these procedures have been published recently because of the
current emphasis on research on the prevention, detection,
and treatment of cancer. Included have been several review
articles. In particular, Krewski and Brown (1981) present a
guide to the statistical literature on carcinogenic risk
assessment which includes an introduction to the issues
discussed in this chapter and an extensive bibliography
grouped by topic. Several references which they list are
also referred to in this paper as particular topics are
discussed, but no attempt is made to provide as complete a
bibliography here. Also, the Interagency Regulatory Liaison
Group (1979) discusses the full range of issues encountered
in carcinogenic risk assessment, including the statistical
issues just introduced for discussion in this chapter.

II. THE ANALYSIS OF LIFETIME TUMOR RATES

A. Site-Specific Tumor Rates

In this section the analysis of site-specific lifetime tumor
rates is discussed (see also Gart *et al*. (1979)) for

experiments with random allocation of animals to two or more
treatment groups. Site-specific rates have been chosen
instead of summary statistics, such as the average number of
tumors per animal or the average number of sites with tumors
per animal, because certain sites might be suspected
carcinogenic targets for a given substance based upon
chemical and biological considerations. Also, the presence
of even one tumor may by itself be life threatening, so an
indicator of the presence of at least one tumor at a given
site is used as opposed to the average number of site
specific tumors per animal.

As an illustration, the statistical analysis of brain
tumor rates in rats fed either a standard diet or a standard
diet supplemented with the sweetener aspartame is presented
in Section II G. References to this illustration are made
throughout this section to provide background. The book by
Stegink and Filer (1983) describes the development of
aspartame and includes a chapter by Cornell, Wolfe and
Sanders (1983) which gives more detail on the statistical
analysis they prepared for presentation to a Public Board of
Inquiry and the Commissioner of the Food and Drug
Administration (FDA).

Much of the research throughout the development of
aspartame was motivated by concern for its carcinogenic
potential even though the ubiquity of the component amino
acids, aspartic acid and phenylalanine, suggests that the
likelihood of carcinogenicity is small. However, from among

the many studies exploring carcinogenic potential by both
in-vitro and in-vivo methods, one study suggested a possible
association of aspartame with an increase in incidence of
brain tumors in rats.  Therefore brain tumors were of
special interest in this illustration.

B.  Data Quality

It is important in controlled experiments that apart from
treatment, conditions be the same for all animals.  It is
also important that the data be accurate and reliable.

Bias, or lack of accuracy, could easily arise with data
on tumor incidence.  The likelihood of finding a tumor
depends not only on the size and location of the tumor, but
also on the number of sections of material examined for each
site.  If more sections were prepared from material from a
given site for rats fed aspartame than for controls, for
instance, then the likelihood of finding a small tumor at
that site among rats fed aspartame would be greater than for
control rats.  A bias would thus have been introduced
through these different examination intensities.  Failure to
"blind" the pathologist with respect to the diets of the
experimental animals could also introduce bias.

Reliability refers to the extent to which data are
reproducable, regardless of its inherent bias.  The use of
different intensities of examination, if not related to
treatment assignment, would lead to a lack of reliability
without introducing bias.  However, a lack of consistent

procedures could lead to observations which are both
inaccurate and unreliable.

The purpose here is not to describe the experimental
procedures used to assure the accuracy and reliability of
the resultant data; instead the purpose is to emphasize the
need for procedures which yield accurate and reliable data.
This need is underscored in the aspartame illustration by
the requirement by the FDA of a second independent review of
the pathological data which was carried out by the
Universities Associated for Research and Education in
Pathology (UAREP) group. The UAREP data are those presented
in the illustration in Section II G. They differ only
slightly from the results of the initial examinations.


C. Multiple Sites

Even though the use of site-specific data has been
suggested, the fact that tumor rates are calculated for
multiple sites should be taken into account in statistical
inference, especially for sites not thought in advance to be
special targets for the substance under study. For
instance, if a substance affects sites independently, and if
the hypothesis of no difference between tumor rates between
treatment and control groups were tested at twenty different
sites with significance level $\alpha=0.05$ for each test, the
expected number of sites at which a significant difference
would be found would be one, even if there were no
differences in carcinogenic effect between treatment and

control. The probability of there being at least one
significant result would be $1-(1-0.05)^{20}=0.64$. Thus, even
if the significance level for each site were 0.05, the
overall probability of rejection of a true null hypothesis
of no carcinogenic treatment effect would be much higher.
To avoid this problem, a lower significance level can be
utilized for each site and then the Bonferroni method would
lead to an acceptably low experiment-wide significance level
for all sites combined. This method is based on an
inequality which shows that an overall experiment-wide
significance level of $\alpha$ or less can be attained by utilizing
significance levels in separate comparisons, in this case
for separate sites, which sum to $\alpha$. For instance, for five
separate sites, a significance level of $\alpha/5$ could be used
for each site. For further discussion of the use of the
Bonferroni inequality in this context (see Gart *et
al.* (1979)).

Such an adjustment in the significance level for each
individual site is not necessarily appropriate. The
multiple testing problem is no doubt less severe than just
indicated because similar sites would likely produce similar
responses to a compound, thus invalidating the independence
assumption. Also, when small numbers of tumors are
observed, it may be computationally impossible to attain a
prescribed significance exactly. If such a test were
declared significant only when the attained significance is
less than a prescribed nominal level, then the result may be

the same as if a higher prescribed level test had been used. Thus, for a prescribed nominal significance, the actual significance level over all sites may be increased only slightly. Fears, Tarone and Chu (1977) consider the case where a nominal significance level of five percent was used for each of many sites and show that the significance level over all sites was elevated to only eight percent. This would call for only a slight reduction in the nominal significance level per site, if any. Use of the much lower Bonferroni critical significance level would mean that a more pronounced carcinogenic effect would have to be produced at an individual exploratory site for the compound to be declared carcinogenic.

Certain patterns of attained significance levels (P values) might also reflect carcinogenic activity. For instance, the P values for several sites might be close to the borderline for significance instead of being distributed uniformly over the unit interval as would be expected in the absence of carcinogenic activity and with independence between sites. This could indicate a carcinogenic activity which affects several sites but does not produce a significant effect at any one site given the size of the experiment. Thus the distribution of P values over exploratory sites should be examined in order to obtain an overall assessment of carcinogenic activity. Mantel (1980) provides a helpful discussion of the issues involved with

respect to multiple sites and suggests alternatives ways of
interpreting P values.

In some cases, likely target sites for a potential
carcinogen can be determined in advance on the basis of
prior knowledge about physiology or chemistry. A Bonferroni
adjustment of significance levels should not be carried out
for such a site even though data are also analysed for many
other sites. Alternatively, there may be some sites which
are singled out in advance as targets of special interest,
while for other sites the experimentation is more
exploratory or a matter of checking to see that there is not
evidence of unusual carcinogenic activity. A substantial
portion of the overall significance level could be allocated
to the special targets sites identified in advance and the
rest of the overall significance level could be allocated to
the remaining sites for an exploratory analysis of P values.

In the example of Section II G, the brain had been
singled out as a likely site for tumor activity before the
hearing of the Public Board of Inquiry. This focus came
only after the data had been available for public scrutiny
for several years. Nevertheless, because this site was of
special concern at the time of the hearing, apart from
others initially investigated, a Bonferroni correction to
take into account testing at multiple sites was not
utilized.

D. Control Data

Another issue is the choice of control data for comparison
of tumor rates. Comparisons with animals not exposed to the
substance of concern are essential. Animals in carcinogenic
experiments have a substantial number of tumors which occur
naturally. This is because animals which have been found to
be sensitive to carcinogens are purposely chosen for such
experiments to minimize the possibility that carcinogenic
activity will fail to be detected. The question arises of
whether or not external, and perhaps historical, control
data should be used since experimental animals of the same
strain may be used in different experiments, perhaps in
different laboratories or at different times. However, sole
reliance on historical controls should be avoided and it is
preferable if an internal control group is included in each
experiment. The propensity for bias in comparisons with
data on external controls is great since the detailed
protocol for the pathological examination of tissue is
unlikely to be the same in each experiment. The laboratory
environment may change from experiment to experiment, as
well. This is particularly true if historical controls are
used since methods for experimental observation, including
the maintenance of a stable environment, are continually
being refined. Thus, in the example of II G, statistical
comparisons are made only with internal controls. Even
though external controls should not form the basis of

detailed comparisons, the results of experiments should be
interpreted in the light of experience on related studies.

E.  Treatment versus Control

It was traditional as a first step in many analyses to
compare each group receiving the compound being tested for
carcinogenicity with a control group.  In some experiments
there may be only one treatment group in addition to
control.  Frequencies of animals with at least one tumor at
a particular site and numbers of animals examined for both
treatment and control can be entered in a standard two by
two table to test the null hypothesis of no increase in
tumor rate due to treatment.  The frequencies of tumors are
usually small enough so that the Fisher-Irwin test can be
used to test the null hypothesis.  If the data in both the
treatment and control group were stratified by variables
such as sex, litter or treatment location, the Mantel-
Haenzsel method could be used for a combined comparison of
rates.  Alternatively, a confidence interval, using the data
in all the strata, could be computed for a parameter such as
the odds ratio or the difference in underlying rates.
Standard procedures for the comparison of rates are
discussed by Fleiss (1981).

Here it suffices to emphasize that it is particularly
important to achieve high power, that is, high probability
of detection of an effect if it exists.  This is facilitated

by the use of one-sided tests, which is appropriate in any
test of toxicity, and by the incorporation of evidence from
different strata into a single test statistic or confidence
interval, as just suggested.

To achieve high power it is important to design an
experiment appropriately. When only two proportions are to
be compared, the paper by Ury and Fleiss (1980) is helpful
in the evaluation of power for sample size determination.
In the next section we will turn our attention to the
analysis of experiments with random assignment to more than
one treatment group. It is not uncommon for animal
experiments to have three treated groups of animals, each
with 100 animals equally divided between males and females
with an assignment of 200 animals to the control group.

F.  Trend Analysis

Multiple comparisons of a control group with each treatment
group allow for a conservative test for a carcinogenic
effect without requiring any assumptions about the
functional form of a possible relationship of carcinogenic
effect to dosage level. This is also a direct procedure for
investigating if particular doses are "no effect" doses.
However, this approach has major drawbacks. It utilizes the
same control group in each comparison, so separate
comparisons of different dosage groups with the control
group do not yield independent results. An adjustment for

multiple testing, such as use of the Bonferroni inequality
discussed earlier, should be used with this approach.

The multiple comparisons method described above has
several drawbacks. The separate results obtained from
different experimental dosage groups may confuse the issue
rather than lead to an overall conclusion. In addition,
patterns of responses over the doses are not taken into
account. Finally, by fractionating the data into several
separate analysis, the power of each test will be
considerably less than that which could be achieved by
simultaneous analysis of data for all levels of dosage.
Thus it is preferable to investigate the possibility of
trends over dosages and not confine the statistical analysis
to separate comparisons of dosage groups with the control
groups.

Methods for analyzing trends in incidence rates have
been described by Armitage (1955) and by Thomas, Breslow and
Gart (1977). Armitage's method consists of the application
of standard weighted least squares calculations to the doses
(or logarithms of doses) as the independent variable and to
the corresponding tumor rates as the dependent variable.
The weight given to each dose and tumor rate is the sample
size upon which the tumor rate is based. The slope estimate
for a simple linear regression model is calculated and is
divided by its estimated standard error to form a standard
normal z statistic. The standard error squared is the
variance estimate $\hat{p}(1-\hat{p})$ divided by the weighted sum of

squares of doses about their mean, where $\hat{p}$ is a pooled
estimate of the overall tumor rate formed by dividing the
total number of animals with tumors by the total number of
animals studied. The test for a trend in tumor rates over
increasing doses presented by Armitage was also given by
Cochran (1954). The z statistic calculated can be squared
to form a chi-square test; the chi-square can in turn be
subtracted from a total variation chi-square statistic to
form a goodness-of-fit test.

Thomas, Breslow and Gart (1977) develop the likelihood
for the same setting but for a model which equates the logit
of the true tumor rate P, given by $[\ell n\ P/(1-P)]$, to a simple
linear regression on dose with slope $\beta$. They develop a test
of the null hypothesis that $\beta=0$ based upon the
hypergeometric distribution. For the comparison of only two
dosages groups this test reduces to the Fisher-Irwin test.
The computer program they have developed will carry out each
of these tests in addition to the test for goodness-of-fit
presented by Armitage. It also will separately print out
the numerator and denominator of the weighted regression
estimate of slope so that they can be summed over strata and
then entered into a ratio for an overall estimate of $\beta$ and
test of the null hypothesis that $\beta=0$.

In the aspartame example in the next section, logistic
regression is applied to obtain an iterative weighted least
squares analysis of tumor incidence rates relative to the
logarithm of dose. It will be seen that the assessment of

rend was crucial to the evaluation of carcinogenicity for aspartame.

## 5. Aspartame Illustration

The illustrative data on brain tumor incidence relative to aspartame ingestion are presented in Table 1 separately for each experiment by sex and type of exposure. The experiments are labelled E-33/34 and E-70. For each cross classification, the number of rats examined and the number found to have brain tumors by UAREP are given. The rats were necropsied at the time of death or sacrifice at about 104 weeks, whichever came first. Their brains were then sectioned and examined microscopically for tumors.

Table 1 Number of Rats with Brain Tumors Detected by UAREP and Total Number of Rats Examined by Exposure Group, Experiment Number and Sex

| Exposure Group | E-33/34 | | | |
| | Males | | Female | |
| | Tumors | Total | Tumors | Total |
|---|---|---|---|---|
| Control | 1 | 59 | 0 | 60 |
| Dose 1 | 1 | 40 | 2 | 40 |
| Dose 2 | 1 | 40 | 0 | 40 |
| Dose 4 | 4 | 40 | 1 | 40 |
| Dose 6-8 | 0 | 40 | 2 | 40 |
| E-70 | | | | |
| Control | 3 | 58 | 1 | 57 |
| Dose 2 | 2 | 39 | 1 | 39 |
| Dose 4 | 1 | 39 | 0 | 40 |

Before turning to data analysis, the experiment leading
to the data in Table 1 will be described more fully.  The
treatment groups within each experiment included two control
groups, one of each sex, which were fed chow with no
aspartame supplement.  Each of these control groups
initially contained sixty rats.

In E-33/34 four groups of 40 rats each were exposed to
aspartame for each sex.   The dosage levels are referred to
as Doses 1, 2, 4 and 6-8 in Table 1, where the dose number
gives the approximate daily dosage of aspartame in 1000
mg/kg.  Aspartame administration began in the diet of
weanling rats.  In E-70 exposure began with the parents and
was continued until death or sacrifice for the experimental
rats.  For each sex two dosage groups were used in addition
to a control group.  The groups which received aspartame are
labelled Dose 2 and Dose 4 to indicate the approximate daily
dosage of aspartame as in E-33-34.

As seen in Table 1, data from brain sections were not
available for a total of 9 of the rats planned for in the
original design.  There was no evidence that the absence of
these data was related to the occurrence of brain tumors.

The procedure used in the analysis reported here was to
first test for the possibility of a dose effect on tumor
rates among rats fed aspartame, not including controls.
Maximum likelihood logistic regression analysis was used for
these calculations based on the logarithm of the dose of
aspartame for the treated rats.  The tumor rates for groups

of rats fed aspartame are given as percentages by experiment
number and sex in Table 2. Males and females were
considered separately and in combination. No significant
trend in tumor incidence over positive dose levels was found
in any of these analyses (P > 0.20 for each analysis).
In addition, the hypothesis that the frequencies of brain
tumors which led to the incidence rates in Table 2 arose as
independent random expressions of the same underlying rate
of brain tumor development, regardless of the dose,
experiment, or sex designation, was tested using the
variance test based on a Poisson distribution (Snedecor and
Cochran (1967)). The resultant chi-square statistic was not
significant (0.25 < P < 0.50). Thus, not only was no trend
observed in the tumor rates but a complete lack of an effect
of varying the level of aspartame ingestion was observed for
rats fed aspartame. Moreover, no experiment or sex effect
was exhibited for rats fed aspartame.

Table 2    Brain Tumor Incidence Rate in Per Cent Detected
           by UAREP for Rats Fed Aspartame by Exposure
           Group, Experiment Number and Sex

| Exposure Group | E-33/34 | | E-70 | |
|---|---|---|---|---|
| | Male | Female | Male | Female |
| Dose 1 | 2.5 | 5.0 | | |
| Dose 2 | 2.5 | 0 | 5.1 | 2.6 |
| Dose 4 | 10.0 | 2.5 | 2.6 | 0 |
| Dose 6-8 | 0 | 5.0 | | |

Because these analyses for rats fed aspartame showed no differential effect of dosage on tumor incidence, the various groups corresponding to different doses of aspartame were pooled into one combined treatment group for further analysis. If a consistent trend had been observed, then all the data on animals fed aspartame and the data on controls would have been incorporated into a dose-response analysis.

Next the data on rats fed aspartame, pooled together because of the lack of a dose-effect relationship, were compared with the data on control rats. The frequency data are presented in Table 3; corresponding tumor rates are given in Table 4 as percentages. In agreement with the analysis of rates displayed in Table 2 for groups of animal fed aspartame, no significant experiment or sex effect exists for these rats.

The tumor rates in the control and combined treatment groups were compared using the Fisher-Irwin exact test. For E-33/34 the rates for the treatment groups did not significantly exceed the rates for the corresponding control groups for either males (P = 0.39) or females (P = 0.20). When the results for males and females are combined using one-sided Mantel-Haenszel statistics, the incidence rate for the treated rats is still not significantly higher than that for the control rats (P > 0.05). Note that even though it was not deemed appropriate to pool the data for different sexes, the results have been incorporated into a single conclusion through use of the Mantel-Haenszel procedure.

Table 3    Number of Rats with Brain Tumors Detected
           by UAREP and Total Number of Rats Examined
           for Control and Combined Aspartame Treat-
           ment Groups by Experiment Number and Sex

| Exposure Group | E-33/34 | | | |
| | Male | | Female | |
| | Tumors | Total | Tumors | Total |
|---|---|---|---|---|
| Control | 1 | 59 | 0 | 60 |
| Treatment | 6 | 160 | 5 | 160 |
| | E-70 | | | |
| Control | 3 | 58 | 1 | 57 |
| Treatment | 3 | 78 | 1 | 79 |

Table 4    Brain Tumor Incidence Rates in Per Cent
           Detected by UAREP for Control and
           Combined Aspartame Treatment Groups by
           Experiment Number and Sex

| Exposure Group | E-33/34 | | E-70 | |
| | Male | Female | Male | Female |
|---|---|---|---|---|
| Control | 1.7 | 0 | 5.2 | 1.8 |
| Treatment | 3.8 | 3.1 | 3.8 | 1.3 |

This same approach will later be used to form a single conclusion for all the data from both experiments.

The data of E-70 give no indication of an effect of aspartame on tumor incidence for either males or females. The P values calculated with the Fisher-Irwin exact test are 0.79 and 0.66 for males and females, respectively. For males the incidence rate for controls is higher than for experimental rats; the two incidence rates are essentially the same for females.

Statistical results for the E-33/34 and E-70 experiments have been combined over the four sex-experiment categories with the one-sided Mantel-Haenszel procedure. This method gives a single summary chi-square statistic from both experiments for the comparison of the tumor rates for control rats and rats treated with aspartame. The resultant statistic is not statistically significant ($P > 0.20$), so the null hypothesis of no treatment effect of aspartame on brain tumor incidence is sustained.

Approximate powers for the E-33/34 and E-70 designs are recorded in Table 5 for several combinations of the tumor rate among control animals ($\pi_1$) and the increase in tumor rate ($\delta$). For example, if the *true* rate for control animals was 0.4% and the *true* rate for dosed animals was 3.4%, then the design of the experiments would lead to finding a significant ($P < 0.05$) difference in *incidence* rates about 77% of the time. Since the true rates are not known, a range of possible values is considered in Table 5. The

Table 5   Power (%) to detect the difference between
          control ($\pi_1$) and dosed ($\pi_2 = \pi_1 + \delta$) tumor rates
          with 240 rats in control group and 480 rats
          in dosed group.  Power was generally less
          than 50% for $\delta$ less than 3%.  One-sided test
          with $\alpha$ = 0.5; E-33/34 and E-70 combined.

| $\pi_1$ | 3% | $\delta$<br>4% | 5% |
|---------|----|----|----|
| 0.4% | 77 | 92 | 98 |
| 1.0% | 66 | 85 | 95 |
| 1.4% | 61 | 81 | 92 |
| 2.0% | 54 | 75 | 88 |

calculations reported in Table 5 are based on sample sizes

for the control and dosed groups of 240 and 480,

respectively.  These sample sizes would result if the data

from the two experiments were combined into just two groups

disregarding sex, experiment number and dose level of

treated animals.  The calculations are based on the fourth

approximation presented by Ury and Fleiss (1980).  The

values in Table 5 are not exact but represent close

approximations to the combined power of the two experiments.

The regulatory conclusion that the null hypothesis of

no aspartame effect should be retained is based upon powers

greater than 50%, the power referred to by the Commissioner

of the FDA (1980) in his discussion of cyclamate for $\delta$ as

small as 3% for a broad range of $\pi_1$ values for controls, and

more specifically, on the considerable higher power of 77%

for $\delta$ = 3% for the small value of $\pi_1$ of 0.4%.

As an alternative to the power calculations just
described after an experiment has been completed, confidence
limits on a difference between treatment and control
populations can be calculated.  If a parameter which
describes this difference has been estimated precisely as
shown by a narrow interval, and if the interval includes a
"no difference" value of the parameter, then this is
evidence that the null hypothesis should be substained.  For
this calculation, experimental and control groups were
compared on the basis of the logarithm of the odds ratio.
This is a statistic which summarizes all of the information
in a four-fold table giving the frequencies of rats with and
without brain tumors separately for the treatment and
control groups.

Based on a 95% confidence interval for the logarithm of
the odds ratio and data from both E-33/34 and E-70, the
tumor rate for rats fed aspartame is no more than 3.4 times
the control rate.  Based on E-70 alone this factor is
reduced to 2.4.

The power and confidence interval calculations, in
conjunction with the hypothesis test results, support the
regulatory decision that aspartame is not carcinogenic.


H.  Clustering by Litter

It seems likely in the aspartame illustration that some of
the animals were from the same litter.  This is particularly
true in E-70 where ingestion of aspartame by parents started

before the birth of the rats entered into the experiment.
However, records on litter membership were not kept.  Litter
effects are common in teratology studies and more recently
have been of concern in carcinogenic studies (see Mantel
(1980)).

Sometimes it is possible to use the litter as  block,
that is, to stratify on litter so that each treatment and
the control are randomly assigned separately to females and
males from that litter.  Only litters of sufficient size and
sex composition could be utilized.  Often exposure is
introduced before birth as in E-70, and litter effects are
necessarily clustered within treatment groups.  When this is
the case, the unit of analysis should be the litter unless
it can be shown that the animals respond independently
within as well as between litters.

Three main approaches have been used for analyses of
litters.

## 1. Analysis of Variance Methods

One approach to the analysis of frequency of response
is to use analysis of variance techniques with the litter as
the unit of analysis.  However, it is not appropriate to
apply these techniques directly to the response frequencies
because of heterogenicity of variances.  If the number of
implants per animal is assumed to be unaffected by treatment
assignment, then the frequencies of response can be assumed
to be binomial random variables and the inverse arcsine
transform can be applied to the observed proportions before

use of the analysis of variance. This only eliminates the effect of varying probabilities of response on the variances. If the litter sizes vary appreciably, a weighted analysis of transformed response proportions could be used. Alternatively, if the litter sizes are large, and the frequencies of animals with tumor per litter small, then these frequencies could be regarded as Poisson variables and the square root transform could be applied before carrying out an analysis of variance. Use of this transform avoids the necessity of a weighted analysis. In actual practice, modifications of these transforms, which avoid problems when all the animals in a litter either respond or fail to respond, are usually employed. These are referenced in a review article by Haseman and Kupper (1979).

Haseman and Kupper (1979) cite examples where binomial and Poissons models do not fit so that transforms based upon these distributions would be inappropriate. In these examples, the variability between litters in the same treatment groups is greater than anticipated with those models. Therefore, other methods have been developed to describe this extra variability and to incorporate it into the analysis.

One of these approaches is due to Cochran (1943) who developed a weighted least squares procedure which provided for different litter sizes and for the estimation of variability of the binomial probability from litter to litter, without the specification of a form for its distribution, and assumed binomial variation within litters. This procedure has been further developed by Kleinman (1973).

## 2. Models for Extra Variability

Two other methods incorporate standard distributions for the variability of model parameters. Williams (1975) develops the beta-binomial model for this setting by assuming that the probability of response per litter in the same group varies in accord with the beta distribution while retaining the assumption of binomial variability within litters. He develops asymptotic likelihood ratio tests for comparing treatment and control groups to see if the same parameters adequately describe the distribution of litter responses in both groups. McCaughran and Arnold (1976) generalize the Poisson distribution in a similar way by assuming that the Poisson mean varies between litters within the same treatment group in a manner described by a gamma distribution. This leads to the negative binomial distribution for litter frequencies of response. They develop transforms based on moment estimates which lead to analysis of variance procedures to be used for comparisons of treatment and control groups of litters.

These models do not allow for negative correlations within litter. Kupper and Haseman (1978) introduced a correlated binomial model which provides a fit to data with extra variability between litters similar to that for the

beta-binomial model, but allows for the alternative of extra
variability within litters, that is, negative intra-litter
correlation.  They also develop  likelihood ratio tests for
comparisons between treatment and control groups.  Examples
of the use of these methods as well as modifications of them
are again cited in the review by Haseman and Kupper (1979).

3.  *Nonparametric Methods*

Differences between the proportion of affected fetuses
in treatment and control groups of litters also can be
assessed without making distributional assumptions by using
standard nonparametric tests, such as the Wilcoxon test.
For more than two groups, comparisons could be made with the
Kruskal-Wallis test or a trend could be investigated with
Jonckheere's test.  Alternatively, the Pitman randomization
test could be applied (see Mantel 1979).

Another approach which is not based on underlying
distributional assumptions is the use of jackknife
methodology.  Gladen (1979) discusses the application of
this to compare treatment and control groups of litters with
a statistic which has an approximate t-distribution.  Butler
(1983) places the overall problem in the framework of
cluster sampling.  This framework enables him to extend
Gladen's method to multiple treatment groups and to develop
an approach for the determination of samples sizes.

III.  TIME TO TUMOR CONSIDERATIONS

When comparing tumor incidence rates among various
experimental groups of animals, it is essential to consider

the effects of animal longevity on tumor production. A group of long-lived animals may appear to have a higher tumor rate than does a group of shorter-lived animals because tumors have had more time in which to develop. Similarly, toxic effects of a substance may lead to shorter lifetimes, thereby masking a carcinogenic effect of the substance because the animals do not live long enough for the tumors to develop. Peto *et al.* (1980) discuss this in more detail and give more examples showing how the relationship between longevity and tumor production can influence the apparent incidence of tumors during an animal's lifetime.

In addition, when comparing tumor production among various animal groups, it may be important to consider times to tumor in addition to the number of animals with tumors because a carcinogenic effect may be manifested by both an increased number of animals with tumors and by a decrease in the time to occurrence of the individual tumors. Thus, evaluation of tumor rates, alone, ignores a potentially important part of the evidence related to carcinogenicity.

Appropriate resolution of these issues depends upon another important consideration, the relationship between the occurrence of the tumor and the time of discovery of the tumor, often at the time of death of the animal. At one extreme, a tumor may be benign and have no impact on the animal's death. Such tumors are discovered only when the animal is sacrificed or dies from other causes. This information allows estimation of the *prevalence* of tumors.

Serial sacrifice experiments are used, primarily, because
they give unbiased information about tumor prevalence. At
the other extreme, a tumor may be discovered near the time
of its occurrence, either because it causes death almost
immediately or because it is palpable or visible. In such
cases, we gain information about the *incidence* of tumors, or
about the incidence of death from tumor. Statistical
methods have been developed to analyze data for either of
these types of tumors, as well as for a mixture of these two
types of tumor, while at the same time addressing the two
issues raised previously, evaluating time to tumor and
controlling for intervening mortality. These methods will
be discussed below. More generally applicable methods are
under development (McKnight and Crowley (1983)).

When analyzing time to tumor data from animal
experiments, there may be information available about the
tumor other than the time of discovery of the tumor. Tumors
may be discovered by palpitation of the animal, by necropsy
at the time of sacrifice, by necropsy at the time of death
from the tumor, or by necropsy at the time of death from
other causes. Such information concerning the mode of
discovery of the tumor may be relevant to the comparison of
tumor rates among animal groups, as discussed above. In
addition, the size, staging, malignancy or multiplicity of
the tumor may be recorded,  and may also be relevant to the
comparison of tumor incidence among animal groups. Such
extra information may be incorporated into survival analysis

methods using methods proposed by Cox (1972), Peto *et al*. (1980), Finkelstein and Wolfe (Chapter 4), or Kodell and Nelson (1980), among others. Krewski and Brown (1981) give references to a variety of other methods proposed for analyzing complex data from animal carcinogenesis experiments. Here, we will review the most standard methods of analysis for such data, which largely ignore such extra information. It is important to recognize that the methods of survival analysis are undergoing rapid development so that methods which are considered standard today may well be supplanted by improved methods in the future.

Most of the statistical methods in use today for the analysis of times to tumor are asymptotic methods. That is, they rely on the central limit theorem to give the approximate distribution of the relevant statistics. An exception to this is the use of small sample permutation tests (Gehan (1965)) for the comparison of survival curves. However, the use of permutation tests for comparing tumor incidence is only appropriate in certain special cases, and some authors have indicated that they should not be used generally for such purposes (Prentice and Marek (1979)).

A.  Incidence

Methods, collectively known as survival analysis methods, have been developed specifically for use with incidence type data while addressing the two major issues mentioned at the

beginning of this section.  For historical reasons we use
the common "survival" terminology even though we are
discussing times to tumor rather than times to death.
Survival analysis methods allow comparisons of times to
tumor among groups of animals while adjusting for
intervening mortality. Such comparisons are often summarized
by survival curves.  In the survival curve for a group of
animals the fraction of animals without tumor is plotted
versus time.  The survival curves are often estimated by the
maximum likelihood method (Kaplan and Meier (1958)) although
particular parametric forms for the survival curves, such as
the Weibull form, may be useful for certain purposes.  The
comparison of survival curves addresses the issue of the
importance of comparing both times to tumor and numbers of
animals with tumor.

The other issue mentioned above, of interceding
mortality when comparing tumor incidence, can be thought of
as a problem of competing risks.  The animals are subject to
the occurrence of several types of risk of death, which are
dichotomized here as tumor or other causes.  These risks of
death compete to cause the death of the animal.  This
problem is addressed by treating the time of death for an
animal without tumor as a right censored observation of the
time to tumor.  An observation of time to tumor for an
animal is called right (left) censored if instead of being
the actual time of tumor it is a lower (upper) bound on the
time at which a tumor would have occurred for the animal.

Time to tumor data from animal experiments generally include information about several types of tumors. As discussed previously in this chapter and reviewed by Gart *et al.* (1979), each type of tumor is often the focus of a separate analysis. The data for time to tumor for one such type of tumor can be denoted as $X_{ij}$, $\delta_{ij}$ for animals $i=1,\ldots,n_j$ and groups $j=1,\ldots,G$, where $X_{ij}$ denotes the time of detection of the tumor of the given type or the time of death for an animal, whichever occurs first, and $\delta_{ij}$ is 1 if a tumor was detected or 0 if the animal died with no tumor detected at time $X_{ij}$. The datum $\delta_{ij}$ is called the censoring indicator for the ij'th animal and equals 0 if the time to tumor for the animal is right censored. From such data, the distinct ordered times of detection of tumor, $T_1 < \ldots < T_K$, can be extracted. Further, let $d_{kj}$ for $k=1,\ldots,K$ and $j=1,\ldots,G$ be the number of tumors detected at time $T_k$ in group j and let $r_{kj}$ be the number of animals alive and with no detected tumor just prior to time $T_k$ in group j. In a more general framework, we say that $d_{kj}$ and $r_{kj}$ record the number of events of interest and the number of animals at risk for the event, respectively, at time $T_k$ in group j.

With the data notation described above, the Kaplan-Meier survival curve estimate at time t for group j can be written as

$$S_j(t) = \prod_{T_k \le t} (r_{kj} - d_{kj})/r_{kj}.$$

Rank tests generalized to be applicable to censored data are often used to compare these survival curves.

Examples of such tests are described in Chapter 4, which
also contains an application to animal carcinogenesis as an
illustration, so the description in this section is brief.
A large class of such tests can be characterized as follows.
Let

$$d_k = \sum_{j=1}^{G} d_{kj} \quad \text{and} \quad r_k = \sum_{j=1}^{G} r_{kj}.$$

Let $\underline{D}_k$ be the column vector $(d_{k1},\ldots,d_{kG})'$ and let $\underline{E}_k$ be the
column vector $(r_{k1},\ldots,r_{kG})'d_k/r_k$ for $k=1,\ldots,K$. Note that
$\underline{D}_k$ and $\underline{E}_k$ are the vectors of the observed and expected
numbers of tumors in each group at time $T_k$, respectively,
assuming equal tumor rates in all the groups. Then the
values $w_k$, $k=1,\ldots,K$, specify the form of a test statistic
vector,

$$\underline{S} = \sum_{k=1}^{K} w_k(\underline{D}_k - \underline{E}_k).$$

If $w_k=1$, the test is the log rank test while if $w_k=r_k$, the
test is Gehan's (1965) generalization of the Wilcoxon test.
The covariance matrix of the vector $\underline{S}$ can be estimated,
using various approximations and depending upon assumptions
about the nature of the variability in the experiment. The
various covariance estimates are discussed in Andersen $et$
$al$. (1982). The covariance estimate based on an assumed
hypergeometric distribution of the observed number of tumors
at each time $T_k$ is commonly used.

B.  Prevalence

A variety of methods have been proposed for analyzing
prevalence data from animal carcinogenesis experiments.
Such data arise when tumors are discovered at times which
are independent, in a statistical sense, of the occurrence
of the tumors.  Dinse and Lagakos (1982) review some of
these methods and propose a methodology of their own.  One
relatively standard approach (Hoel and Walburg (1972))
consists of grouping the data according to intervals along
the time axis.  For each interval, the animals sacrificed or
dying during the interval are classified into a 2 by G table
according to group and whether or not a tumor was found.
Evidence comparing the tumor prevalence rates among groups
is then combined across intervals using the Mantel-Haenszel
method.  This yields a vector statistic, $\underline{S}$, the sum across
intervals of differences between the observed and expected
number of tumors in the groups, assuming the null hypothesis
of no differences among groups in tumor prevalence rates.
Peto *et al.* (1980) propose a combination of the log rank
statistic and the prevalence statistic for use with data
where tumors have been classified as to whether they were
the cause of death or were incidental to the death of the
animal.

C.  Dose Response

As discussed previously in this chapter, the evaluation of
data from animal experiments for tumorogenicity should

generally be directed towards discovery of a dose-response
relationship between the level of exposure to a substance
and the incidence of tumors.  The methods for incidence and
prevalence data, discussed previously, are based on a
statistic which is a vector $\underline{S}$, the vector based on
differences between the observed number of tumors and the
expected number of tumors in each group.  A large sample
estimate of the covariance matrix, $\ddagger$, of the components of $\underline{S}$
can also be computed.  This allows a regression type of
analysis to be carried out to evaluate the association
between the exposure level and the observed minus expected
discrepancies.  Letting $\underline{C}$ be the vector of exposure levels,
the test statistic is $\underline{C}'\underline{S}$ with variance $\underline{C}'\ddagger\underline{C}$.

An alternative approach can also be used to evaluate
evidence for a dose response relationship based on incidence
type data.  The Cox proportional hazards regression model
(Cox (1972) and Kalbfleisch and Prentice  (1980)) can be
used to model the association between time to tumor and
exposure level.  The Cox model methodology is particularly
useful in situations where there are differences in animal
characteristics among the experimental groups for which
statistical adjustment is to be made.  In such cases, those
characteristics can be included along with indicator
variables for groups membership as covariates in the Cox
model.  This could allow, for example, an adjustment to be
made for differences between groups in animal weight at
birth, if such an adjustment were appropriate from
physiological considerations.

IV.  QUANTAL RESPONSE LOW DOSE EXTRAPOLATION

The identification of a substance as a carcinogen in a
species of animal leads to the question of whether the
substance is also a human carcinogen.  However, reliable
epidemiologic data on the long term follow-up of people
receiving varying degrees of exposure are typically not
available.  Similarly, insufficient information on the
mechanisms and rates of metabolization of the substance may
make suspect a decision based solely on biochemical and
physiological principles.  It is current practice, then, for
animal carcinogens also to be considered human carcinogens
unless there is substantial evidence to the contrary.  Of
course, as more information becomes available the labelling
of a substance as a human carcinogen may be qualified or
removed.

A substance identified as an animal carcinogen may be
totally restricted for some human uses, for example as a
food additive (Public Law 85-929, 85th Congress, H.R. 13254,
September 6, 1958).  It may still be a candidate for other
human uses, however, if exposure levels associated with no
or only minimally increased risk could be maintained.  The
establishment of human dose-response relationships required
for the latter situation is usually based, at least
partially, on the results of animal carcinogenicity
experiments.  However, the dose-response relationship in
animals for extremely low risks is not addressed in the
standard animal carcinogenicity trial.  In fact, the number

of animals required to detect increased risks of the
magnitude considered here precludes their direct observation
(Mantel and Byran, 1961). Researchers thus often resort to
assumed functional forms of dose-response relationships in
order to extrapolate to the lower levels of risks which are
usually associated with extremely low doses. Technically,
this is interpolation, since a control group is typically
included in most experiments, though here it will be
referred to as extrapolation.

Several of the more commonly used low dose
extrapolation models for quantal response data are presented
and discussed in Section IV A. This is followed in Section
IV B with a discussion of the methods to accommodate
spontaneous cancer rates, incorporate pharmocokinetics and
extrapolate from animals to man. A summary and discussion
is given in Section IV C. Other recently published reviews
which discuss general problems of low-dose extrapolation
include Crump (1979), Armitage (1982) and Hogan and Hoel
(1982). The bibliography prepared by Krewski and Brown
(1981) includes many relevant references.

A few general comments on low-dose extrapolation should
be kept in mind when assessing the usefulness of any of the
methods discussed below. For example, none of the models
considered here allows for a population threshold in dose
below which no increase in risk exists for *any* individual.
Though kinetic models can be developed which allow for
thresholds for individuals, the existence of thresholds for

a population does not necessarily follow (Cornfield *et al.*
(1978)). Many of the models currently used allow for
individual thresholds which here will be referred to as
tolerances. Since exposure to a carcinogen at even the
lowest doses may be associated with some increase in risk in
the population, no absolutely safe dose is assumed to exist.
Extrapolation is thus made to virtually safe doses, VSD's,
associated with extremely low risks, say $10^{-8}$ or $10^{-6}$. The
magnitude of this risk reflects the upper limit on the
"acceptable" increase in the number of cancer cases due to a
lifetime of exposure.

Extrapolation models should not, and usually cannot, be
selected solely on the closeness of their fits to specific
data sets. Though several models may provide quite good
fits, their predicted risks at extremely low doses may
differ by several orders of magnitude. Current
understanding of carcinogenic processes in general and
characteristics of the specific exposures and organ sites
involved should influence the selection of extrapolation
models.

Since there are obvious undesirable costs associated
with underestimating the risk associated with specific doses
of a substance, a conservative interpretation of the fitted
dose-relationship is used. The conservatism is usually
introduced by extrapolating from the upper confidence limit
for the dose-response relationship or from some other
function which is assumed to be an upper bound for it.

However, too conservative an approach can result in an imbalance between the risks and benefits of the use of the substance (Cornfield (1977)). As more information becomes available, these estimates can, of course, be updated.


A.  Specific Models

1.  *Mantel - Bryan*

    The Mantel-Bryan model (Mantel and Bryan (1961); Mantel *et al.* (1975)) assumes each animal possesses a tolerance to exposure below which it will not develop a tumor. These tolerances are assumed to be approximately log normally distributed across animals. The problem is then to determine the dose, d, which exceeds the tolerance of only a small percentage of the animals. Using Abbott's formula (Abbott (1925)) to incorporate a spontaneous tumor rate, the risk of cancer at dose d can be written

$$P(d) = P(0) + (1-P(0)) A(d)$$

where      $A(d) = [P(d)-P(0)]/[1-P(0)] = \phi(\alpha+\beta \log_{10}(d))$,

P(0) is the spontaneous tumor rate, $\alpha$ and $\beta$ are parameters of the tolerance distribution and $\phi$ (·) denotes the standard normal cumulative function. Mantel and Bryan (1961) and Mantel *et al.* (1975) argue that sufficient information may not always be available to estimate $\beta$ reliably. Based on empirical evidence, however, they suggested that setting $\beta$ equal to one would lead to conservative estimates of the true VSD. Thus, only the two parameters P(0) and $\alpha$ need to be estimated. The extrapolation is performed from the upper

confidence limit of $\alpha$ so as to guard against over estimating
the virtually safe dose.  These authors also suggest that
the data from the higher doses may be excluded from the
estimation of the VSD so as not to discourage
experimentation at high doses.

The major criticisms of this model include the reliance
on the assumption of a slope of unity and the absence of an
underlying carcinogenic mechanism.  Krewski and Van Ryzin's
(1981) examination of 20 substances shows that the estimated
slope is less than 1.0 in four situations and between 1.0
and 1.4 in an additional 7.  The consistent conservatism of
the model can thus be questioned.  In addition, even if
empirical evidence suggests the slope is generally less than
1.0 in the interval of observation, the slope at extremely
low doses may be lower.  Of course, this assumption can be
relaxed and the slope estimated by maximum likelihood.  The
probit ($\beta$ estimated) and Mantel-Bryan ($\beta=1$) models are,
however, always concave at low risks while other models
allow for linear or convex relationships.  Relative to the
other models, the Mantel-Bryan method may not be as
conservative as was originally expected.

## 2. Multistage

The multistage model is based on the assumption that
cancers originate from a single cell which has progressed
through k irreversible changes which are usually referred to
as stages or "hits".  The rate, $\lambda_i$, of the progression from
stage i-1 to i, is assumed to be independent of age.  The

age-specific cancer incidence rate is then given by
(Armitage and Doll (1954))

$$I(t) = t^{k-1}(\prod_{i=1}^{k} \lambda_i)/(k-1)!$$

and the probability of tumor by age T is given by

$$P(T) = 1-\exp\ [-Q(T)\cdot(\prod_{i=1}^{k} \lambda_i)/(k-1)!],$$

where $Q(T)$ does not depend on the $\lambda_i$'s. If the rates
$\lambda_i = \alpha_i + \beta_i d$ are allowed to depend on dose d in a linear
fashion with $\alpha_i \geq 0$ and $\beta_i \geq 0$, and T is taken to be the natural
life expectancy of the animal, then the probability of tumor
in an animal exposed to a dose of d for its entire life is

$$P(d) = 1-\exp\ [-Q(T)\prod_{i=1}^{k} (\alpha_i + \beta_i d)/(k-1)!]$$

$$= 1-\exp\ (-\sum_{i=0}^{k} q_i d^i).$$

The single hit model with k=1 is a special case of the
above. The coefficients $q_i$ are typically estimated by
maximum likelihood subject to the constraint $q_i \geq 0$ (Crump
et al. (1977); Guess et al. (1977)). Implicit in this model
is the treatment of a spontaneous tumor rate with Abbott's
formula with

$$A(d) = 1-\exp\ (-\sum_{i=1}^{k} q_i d^i).$$

For extremely low values of d,

$$\log\ (A(d)) \sim \log\ (q_\lambda) + \lambda\ \log\ (d),$$

where $\lambda$ denotes the lowest power of dose with a nonzero coefficient. Note that even though the maximum likelihood estimate (MLE) for $q_1$ may equal 0, the upper confidence limit for $q_1$ will typically be greater than zero. Thus, at extremely low doses, the upper confidence limit of risk will be approximately linear with unit slope on the log-log scale.

The linearity at low doses results in lower VSD's from the multistage model than from several other models. Krewski and Van Ryzin (1981) show this to be the case for 19 of 20 data sets considered. The exception was vinyl chloride which is the only substance exhibiting a strongly concave dose-response relationship.

The theoretical foundation of the multistage model has also been utilized to establish useful epidemiologic research and analysis strategies (Brown and Chu (1982)). Supporting evidence indicates that the multistage model may be a reasonable extrapolation model (Peto (1977)). It must be remembered, however, that it is but one of many mathematical models proposed to describe the carcinogenic process and that the appropriateness of the multistage model is unlikely to be universal (Whittemore and Keller (1978)).

The multistage model has been generalized to incorporate time to tumor data (Hartley and Sielken (1977)). This extension leads to alternative methods of measuring the increased risk of cancer associated with exposure to a given dose.

3. *Gamma Multihit*

Rai and Van Ryzin (1979) have developed a model based on earlier work by Cornfield (1954) which is also based on "hits" but assumes the number of hits follows a Poisson distribution with parameter $\theta d$. The probability that a cell has experienced k or more hits by time t is

$$P(d) = \sum_{i=k}^{\infty} (\theta d)^i e^{-\theta d}/i!$$

$$= \int_0^d [\theta^k t^{k-1} \exp(-\theta t)/\Gamma(k)]dt$$

where $\theta > 0$, $k > 0$ and $\Gamma(k)$ is the complete gamma function. Rai and Van Ryzin (1979) propose that a spontaneous tumor rate be incorporated in the model using Abbott's formula with

$$A(d) = \int_0^d [\theta^k t^{k-1} \exp(-\theta t)/\Gamma(k)]dt.$$

The estimates for k and $\theta$ are obtained by maximum likelihood and extrapolation is performed, as before, from the upper confidence limit for $\theta$. The multihit model has a mechanistic basis, like the multistage model, but can also be represented in terms of a distribution of tolerances, like the Mantel-Bryan model.

Haseman *et al.* (1981) have identified several concerns in the use of the gamma multihit model. These include the possibility of extreme estimates of VSD, either high or low; unrealistically narrow confidence intervals; and the possibility of unreasonable estimates for k, the number of hits. Some of these concerns have been partially addressed (Van Ryzin and Rai (1980, 1981)) though the reliability of VSD estimates from this model has yet to be established.

## 4. Linear

Hoel *et al.* (1975) present a conceptually and computationally simple extrapolation procedure which assumes only that the dose-response relationship is convex at low doses. Consider an experiment conducted at a single dose, $d^*$, with the observed proportion of animals with tumor denoted by $\hat{p}(d^*)$. If a convex dose response relationship holds for $0<d<d^*$ and $P(0)=0$, then the straight line through the origin

$$P(d) = d[\hat{p}(d^*)/d^*]$$

provides an upper bound for the true dose-response relationship at the low doses. Conservative estimates for the VSD can then be obtained from this straight line. Variations of this very simple model include incorporation of a spontaneous tumor rate (Hoel *et al.* (1975)), rules for inclusion of several experiment doses and adjustments to account for uncertainty in point estimates (Gaylor and Kodell (1980)). Van Ryzin (1980) and Gaylor and Kodell (1980) have suggested considering linear extrapolation from doses and fitted risks obtained from other assumed functional dose response relationships. For example, the multistage model could be fit to the experimental data and the VSD for a risk of $10^{-2}$ obtained. VSD's for lower risks could then be obtained by linear extrapolation between this point and the origin. The risk from which linear extrapolation is performed could vary with the amount of information in the data set. Recognizing that many of the

extrapolation models are quite similar in the regions of
observation, the inclusion of linear extrapolation would
lead to a greater homogeneity of VSD's among models.
Linearity at low doses is also consistent with the
multistage model. In fact, the VSD's from linear
extrapolation and the multistage model are usually quite
similar.

B.   Issues in Interpretation and Application

1.   *Incorporation of Background Rates*

Several of the models discussed above incorporate
background tumor rates using Abbott's formula. This
adjustment implies independence of the background and dose-
related incidences with $1-P(d) = [1-P(0)][1-A(d)]$ where $A(d)$
is the tolerance function. Alternatively, additive
background and dose-related incidences can be represented as
$P(d)=F(d_0+d)$ where F can range between 0 and 1. Note that
the multistage model is additive yet naturally results in
the use of Abbott's formula to adjust for the spontaneous
tumor rate.

The low-dose behavior of additive and independence
models differ substantially. Crump *et al*. (1976) have shown
that if the background were additive then the excess risk
will be approximately linear at low doses. Hoel (1980) has
shown that even if as low as 1% of the background were
additive then the dose response relationship would be
approximately linear at low doses. Haseman *et al*. (1981)

have shown the same for the gamma multihit model.  This is
cause for concern for the assumed "conservativeness" of
models which are not linear at the low doses.  Biological
validity has not yet been established, however, for either
of these methods of incorporating a spontaneous tumor rate.

## 2.  Simple Pharmacokinetics

It is now known that some chemicals are carcinogenic
only after they have been metabolized into a form that is
capable of interaction with DNA.  Since nonlinear
pharmocokinetics many be involved in the transformation of
the chemical into its metabolites, the use of administered
dose in the extrapolation models may not reflect the true
level of the carcinogenic substance.  Instead, tumor
response should be related to a measure of metabolite
concentration in the target cell.  Cornfield (1977)
discusses the incorporation of a simple kinetic model which
allows for the toxic substance to be activated and
deactivated in separate and simultaneous reactions.  Gehring
and Blau (1977) have developed a pharmocokinetic model using
vinyl chloride which allows for activation, detoxification
and repair, which are assumed to follow Michaelis-Menten
kinetics.  Anderson et al. (1980) have demonstrated that
ignoring the nonlinear kinetics can result in an
underestimate of the VSD, the magnitude depending on the
specific model used.  In addition, Hoel et al. (1983)
conclude that observed nonlinearities in dose response

curves may, in fact, reflect nonproportional relationships between administered and effective doses.

## 3. *Animal-to-man Extrapolation*

In addition to the difficulties involved in the extrapolation from high to low doses in animals, there are also difficulties involved in the extrapolation from animals to humans. Some biological basis for the extrapolation rule is needed which, as before, may depend on specific substance/organ site combinations and different manners and rates of metabolism.

Currently, extrapolation is based on an assumed comparability across species of specific units of measurement of dose. For example, the National Academy of Sciences (1975) has proposed as a "working hypothesis" that the carcinogenic potential of the consumption of one milligram of substance per kilogram of body weight per day (mg/kg/day) is constant across species. If experimental doses are measured in mg/kg/day then the VSD for mice, $VSD_M$, would equal the VSD for humans, $VSD_H$. The Environmental Protection Agency (1980) extrapolates across species by assuming that the doses measured in mg/surface area/day are comparable across species. Mantel and Schneiderman (1975) discuss this rule which is also supported by the correspondence of toxicity to anticancer drugs across species found by Freireich *et al.* (1966). Surface area can be approximated by weight raised to the power 2/3. When $VSD_M$ is measured in mg/kg/day, the $VSD_H$ can be obtained by the approximation

$$(0.03)^{1/3} VSD_M = (70)^{1/3} VSD_H$$

taking the average weight of a mouse as 0.03 kg and the average weight of a man as 70 kg. When $VSD_M$ is given in ppm, the average mg intake per day is usually assumed to be given by mg=ppm·f·w where f is the fraction of body weight consumed per day as food and w is body weight. Empirical values for the feeding factor, f, are 0.028, 0.05 and 0.13 for man, rats and mice, respectively (Environmental Protection Agency (1980)). Adjustment to a $VSD_H$ can then be made using the above formula.

## C.  Summary and Discussion

Extrapolation from high to low doses in animals and from laboratory animals to humans are the two major components of low-dose extrapolation. The relative strengths and weaknesses of various approaches and models for each component have been examined. Currently, there is no extrapolation algorithm which is universally preferred or recognized as consistently accurate.

Despite the uncertainty of estimates of VSD's, low-dose extrapolation and risk assessment will continue to be performed. The need to prevent human exposure to carcinogens at doses associated with high levels of risk is, of course, the prime reason. The imperfect nature of the information obtained from low-dose extrapolation should be recognized and results interpreted accordingly (Munro and

WOLFE AND BUTLER

Krewski (1981); Calkins *et al.* (1980)). Several
extrapolation models and animal-to-man conversions should
always be considered. Predicted levels of risks, as opposed
to their upper limits, and assessment of goodness-of-fit of
the different models should also be considered (Brown
(1978)). The estimated VSD's should be considered as inputs
to the risk assessment process which are to be incorporated
with other information from epidemiology, toxicology,
physiology and biochemistry.

REFERENCES

Abbott, W.S. (1925). Method of computing the effectiveness
of an insecticide. *Journal of Economic Entomology*,
*18*: 265-267.

Andersen, P.K., Borgun, O., and Gill, R. (1982). Linear
nonparametric tests for comparison of counting
processes, with applications to censored survival
data. *International Statistical Review*, 50:219-258.

Anderson, M.W., Hoel, D.G., and Kaplan, N.L. (1980). A
general scheme for the incorporation of
pharmocokinetics in low-dose risk estimation for
chemical carcinogenesis: Example-vinyl chloride.
*Toxicology and Applied Pharmacology*, 559: 154-161.

Armitage, P. (1955). Tests for linear trends in proportions
and frequencies. *Biometrics*, *11*: 375-386.

Armitage, P. (1982). The assessment of low-dose
carcinogenicity. *Biometrics*, *28* (*Supplement*):
119-129.

Armitage, P. and Doll, R. (1954). The age distribution of
cancer and a multi-stage theory of carcinogenesis.
*British Journal of Cancer*, *8*: 1-12.

Brown C.C. and Chu K.C. (1982). Approaches to epidemiologic
analysis of prospective and retrospective studies:
example of lung cancer and exposure to arsenic. In
*Environmental Epidemiology: Risk Assessment* (Ross
L. Prentice and Alice S. Whittemore, eds.). Siam,
Philadelphia, pp. 94-106.

Brown, C. (1978). Statistical aspects of extrapolation of
    dichotomous dose response data. *Journal of the
    National Cancer Institute*, *60*: 101-108.

Butler, W.J. (1983). *Cluster Sample Analysis for
    Teratogenicity Studies*. Technical Report,
    Department of Biostatistics, University of Michigan.

Calkins, D.R., Dixon, R.L., Gerber, C.R., Zarin, D., and
    Omenn, G.S. (1980). Identification,
    characterization, and control of potential human
    carcinogens: A framework for federal decision
    making. *Journal of the National Cancer Institute*,
    *64*: 169-176.

Cochran, W.G. (1943). Analysis of variance for percentages
    based on unequal numbers. *Journal of the American
    Statistics Association*, *38*: 287-301.

Cochran, W.G. (1954). Some methods of strengthening the
    common chi-square tests. *Biometrics,10*: 417-451.

Commissioner, Food and Drug Administration (1980).
    *Cyclamate (Cyclamic Acid, Calcium Cyclamate, and
    Sodium Cyclamate); Commissioner's Decision*.
    Department of Health and Human Services, Food and
    Drug Administration (Docket No. 76F-0392), Federal
    Register, Vol. 45, No. 181, Tuesday, September 16,
    1980, pp. 61474 ff.

Cornell, R.G., Wolfe, R.A., and Sanders, P.G. (1983).
    Statistical Issues. In *Aspartame: Advances in
    Biochemistry and Physiology* (L.D. Stegink and
    L.J. Filer, Jr., eds.). In press.

Cornfield, J. (1954). Measurement and comparison of
    toxicity: the quantal response. In *Statistics and
    Mathematics in Biology* (O. Kempthorne, ed.). Iowa
    State College Press, Ames, Iowa, pp. 327-344.

Cornfield, J. (1977). Carcinogenic risk assessment.
    *Science, 198*: 693-699.

Cornfield, J., Carlborg, F.W., and Van Ryzin, J. (1978).
    Setting tolerances on the basis of mathematical
    treatment of dose-response data extrapolated to low
    doses. *Proceeding of the First International
    Congress on Toxicology* (G.L. Plaa and W.A.M. Duncan,
    eds.). Academic Press, New York, pp. 143-163.

Cox, D.R. (1972). Regression models and life tables.
    *Journal of the Royal Statistical Society* B, *34*:
    187-220.

468                                    CORNELL, WOLFE AND BUTLER

Crump, K.S., Hoel, D., Langley, C., and Peto, R. (1976).
Fundamental carcinogenic processes and their
implications for low-dose risk assessment. *Cancer
Research, 36:* 2973-2979.

Crump, K.S., Guess, H., and Deal, K. (1977). Confidence
intervals and test of hypothesis concerning dose
response relations inferred from animal
carcinogenicity data. *Biometrics, 33:* 437-451.

Crump, K.S. (1979). Dose response problems in
carcinogenesis. *Biometrics, 35:* 157-167.

Dinse, G.E. and Lagakos, S.W. (1982). Nonparametric
estimation of lifetime and disease distributions
from incomplete observations. *Biometrics, 38:*
921-932.

Environmental Protection Agency (1980). Water Quality
Criteria Documents. *Federal Register, 45:*
79318-79379.

Fears, T., Tarone, R., and Chu, K. (1977). False-positive
and false-negative rates for carcinogenicity
screens. *Cancer Research, 37:* 1941-1945.

Fleiss, J.L. (1981). *Statistical Methods for Rates and
Proportions, Second Edition.* John Wiley and Sons,
New York.

Freireich, E.J., Gehan, E.A., Rall, D.P., Schmidt, L.H., and
Skipper, H.E. (1966). Quantitative comparison of
toxicity of anticancer agents in mouse, rat,
hamster, monkey and man. *Cancer Chemotherapy
Reports, 50:* 219-243.

Gaylor, D.W. and Kodell, R.L. (1980). Linear extrapolation
algorithm for low dose risk assessment of toxic
substances. *Journal of Environmental Pathology and
Environmental Health, 4:* 305-312.

Gart, J.J., Chu, K.C., and Tarone, R.E. (1979). Statistical
issues in the interpretation of chronic bioassay
tests for carcinogenicity. *Journal of the National
Cancer Institute, 62:* 957-974.

Gaylor, D.W. and Shapiro, R.E. (1979). Extrapolation and
risk estimation for carcinogenesis. In *Advances in
Modern Technology, Vol. 1: New Concepts in Safety
Evaluation (Part 2)* (M.A. Mehlman, R.E. Shapiro, and
H. Blumenthal, eds.). Wiley, New York, pp. 65-87.

Gehan, E.A. (1965). A generalized Wilcoxon test for
comparing arbitrarily singly censored samples.
*Biometrika, 52:* 203-223.

Gehring, P.J. and Blau, G.E. (1977). Mechanisms of carcinogenesis: Dose response. *Journal of Environmental Pathology and Toxicology*, 1: 163-179.

Gladen, B. (1979). The use of the jackknife to estimate proportions from toxicological data in the presence of litter effects. *Journal of the American Statistical Association*, 74: 278-283.

Guess, H.A., Crump, K.S., and Peto, R. (1977). Uncertainty estimates for low-dose extrapolation of animal carcinogenicity data. *Cancer Research*, 37: 3475-3483.

Hartley, H.O. and Sielken, R.L. (1977). Estimation of "safe doses" in carcinogenic experiments. *Biometrics*, 33: 1-30.

Haseman, J.K., Hoel, D.G., and Jennrich, R.I. (1981). Some practical problems arising from use of the gamma multi-hit model for risk estimation. *Journal of Toxicology and Environmental Health*, 8: 379-386.

Haseman, J.K. and Kupper, L.L. (1979). Analysis of dichotomous data from certain toxicological experiments. *Biometrics*, 35: 281-293.

Hoel, D.G. (1972). A representation of mortality data by competing risks. *Biometrics*, 28: 475-488.

Hoel, D.G. (1979). Animal experimentation and its relevance to man. *Environmental Health Perspectives*, 32: 25-30.

Hoel, D.G. (1980). Incorporation of background response in dose-response models. *Federation Proceedings*, 39: 67-69.

Hoel, D.G., and Walburg, H.E. (1972). Statistical analysis of survival experiments. *Journal of the National Cancer Institute*, 49:361-372.

Hoel, D.G., Gaylor, D., Kirschstein, R., Saffioti, U., and Schneiderman, M. (1975). Estimation of risks of irreversible delayed toxicity. *Journal of Toxicology and Environmental Health*, 1: 133-151.

Hoel, D.G., Kaplan, N.L., and Anderson, M.W. (1983). Implication of nonlinear kinetics or risk estimation in carcinogenesis. *Science*, 219: 1032-1037.

Hogan, M.D. and Hoel, D.G. (1982). Extrapolation to man. In *Methods in Toxicology* (A. Wallace Hayes, ed.). Raven, New York, pp. 711-731.

Interagency Regulatory Liaison Group (1979). Scientific
    bases for identification of potential carcinogens
    and estimation of risks. *Journal of the National
    Cancer Institute*, 63: 244-268.

Kalbfleisch, J.D. and Prentice, R.L. (1980). *The
    Statistical Analysis of Failure Time Data*. Wiley,
    New York.

Kaplan, E.L. and Meier, P. (1958). Nonparametric estimation
    from incomplete observations. *Journal of the
    American Statistical Association*, 53: 475-481.

Kodell, R.L. and Nelson, C.J. (1980). An illness-death
    model for the study of the carcinogenic process
    using surveillance data. *Biometrics*, 36: 267-277.

Kodell, R.L., Shaw, G.W., and Johnson, A.M. (1982).
    Nonparametric joint estimators for disease
    resistance and survival functions in survival/
    sacrifice experiments. *Biometrics*, 38: 43-58.

Kleinman, J.C. (1973). Proportions with extraneous
    variance: single and independent samples. *Journal
    of the American Statistical Association*, 68: 46-54.

Krewski, D. and Brown, C. (1981). Carcinogenic risk
    assessment: a guide to the literature. *Biometrics*,
    37: 353-366.

Krewski, D. and Kovar, J. (1982). Low dose extrapolation
    under single parameter dose response models.
    *Communications in Statistics, Series B*, 11: 27-46.

Krewski, D. and Van Ryzin, J. (1981). Dose response models
    for quantal response toxicity data. In *Statistics
    and Related Topics* (M. Csorgo, D. Dawson, J.N.K. Rai
    and E. Saleh eds.). North-Holland, New York,
    pp. 201-231.

Kupper, L.L. and Haseman, J.K. (1978). The use of a
    correlated binomial model for the analysis of
    certain toxicological experiments. *Biometrics*, 34:
    69-76.

Mantel, N. (1979). Ridit analysis and related ranking
    procedures-use at your own risk. *American Journal
    of Epidemiology*, 109: 25-29.

Mantel, N. (1980). Assessing laboratory evidence for
    neoplastic activity. *Biometrics*, 36: 381-399.

Mantel, N. and Bryan, W. (1961). 'Safety' testing of
    carcinogenic agents. *Journal of the National Cancer
    Institute*, 27: 455-470.

Mantel, N., Bohidar, H., Brown, C., Ciminear, J., and Tukey,
    J. (1975). An improved Mantel-Bryan procedure for
    'safety' testing of carcinogens. *Cancer Research*,
    35: 865-872.

Mantel, N. and Scheiderman, M. (1975). Estimating "safe"
    levels, a hazardous undertaking. *Cancer Research*,
    359: 1379-1386.

McCaughran, D.A. and Arnold, D.W. (1976). Statistical
    models for numbers of implantation sites and
    embryonic deaths in mice. *Toxicology and Applied
    Pharmacology*, 38: 325-333.

McKnight, B. and Crowley, J. (1983). *Tests for Differences
    in Tumor Incidence Based on Animal Carcinogenesis
    Experiments*. Technical Report No. 56, Department of
    Biostatistics, University of Washington.

Munro, I. and Krewski, D. (1981). Risk assessment and
    regulatory decision making. *Food and Cosmetics
    Toxicology*, 19: 549-560.

National Academy of Sciences (1975). Carcinogenesis in man
    and laboratory animals. In: *Pest Control: An
    Assessment of Present and Alternative Technologies.
    Vol 1. Contemporary Pest Control Practices and
    Prospects*. National Academy of Sciences, Washington
    D.C., pp. 66-82.

Peto, R. (1977). Epidemiology, multistage models and short-
    term mutagenicity tests. In *Origins of Human
    Cancer, Book C: Human Risk Assessment* (H.H Hiatt,
    J.W. Watson, and J.A. Winston, eds.). Cold Springs
    Harbor Laboratory, Cold Springs Harbor,
    pp. 1403-1428.

Peto, R., Pike, M.C., Day, N.E., Gray, R.G., Lee, P.N.,
    Parish, S., Peto, J., Richards, S., and Wahrendorf,
    J. (1980). Guidelines for simple, sensitive
    significance tests for carcinogenic effects in long-
    term animal experiments. In *Long-Term and Short-
    Term Screening Assays for Carcinogens: A Critical
    Appraisal*. IARC Monographs on the Evaluation of the
    Carcinogenic Risk of Chemicals to Humans, Annex to
    Supplement 2, International Agency for Research on
    Cancer, Lyon, pp. 311-426.

Prentice, R.L. and Marek, P. (1979). A qualitative
    discrepancy between censored data rank tests.
    Biometrics, 35: 861-867.

Public Board of Inquiry (1980). Aspartame: Decision of the
    Public Board of Inquiry. Department of Health and
    Human Services, Food and Drug Administration (Docket
    No. 75F-0355), Washington D.C.

Rai, K. and Van Ryzin, J. (1979). Risk assessment of toxic
    environmental substances using a generalized multi-
    hit dose response model. In Energy and Health
    (N. Breslow and A. Whittemore, eds.). SIAM,
    Philadelphia, pp. 99-117.

Snedecor, G.W. and Cochran, W.G. (1967). Statistical
    Methods, 6th Edition. Iowa State University Press,
    Ames, Iowa, p. 232.

Steglink, L.D. and Filer, L.J. (1983). Aspartame: Advances
    in Biochemistry and Physiology. In press.

Thomas, D.G., Breslow, N., and Gart. J. (1977). Trend and
    homogeneity analyses of proportions and life table
    data. Computers and Biomedical Research, 10:
    373-381.

Ury, H.K. and Fleiss, J.L. (1980). On approximate sample
    sizes for comparing two independent proportions with
    the use of Yates' correction. Biometrics, 36:
    347-352.

Van Ryzin, J. (1980). Quantitative risk assessment.
    Journal of Occupational Medicine, 22: 321-326.

Van Ryzin, J. and Rai, K. (1980). The use of quantal
    response data to make predictions. In The
    Scientific Basis of Toxicity Assessment
    (H.R. Witschi, ed.). Elsevier/North-Holland
    Biomedical Press, New York, pp. 273-290.

Van Ryzin, J. and Rai, K. (1981). A generalized multi-hit
    dose-response model for low-dose extrapolation.
    Biometrics, 37: 341-352.

Whittemore, A. and Keller, J.B. (1978). Quantitative
    theories of carcinogenesis. SIAM Review, 20: 1-30.

Williams, D.A. (1975). The analysis of binary response from
    toxicological experiments involving reproduction and
    teratogenicity. Biometrics, 31: 949-952.